COMPARATIVE
PLANT ANATOMY

These Studies are designed to inform the mature student—the undergraduate upperclassman and the beginning graduate student—of the outstanding advances made in various areas of modern biology. The books will not be treatises but rather will briefly summarize significant information in a given field and interpret it in terms of our current knowledge of the rapidly expanding research findings within the life sciences. Also it is hoped that the Studies will be of interest to teachers and research workers.

BIOLOGY ←
STUDIES

Sherwin Carlquist
Claremont Graduate School

COMPARATIVE
PLANT ANATOMY

**A GUIDE TO TAXONOMIC AND
EVOLUTIONARY APPLICATION OF
ANATOMICAL DATA IN ANGIOSPERMS**

Holt, Rinehart
and Winston
New York

foreword ▶▶▶▶▶▶

    The purposes of this book are as follows: (1) to provide a guide to literature on comparative anatomy of angiosperms, organized according to anatomical categories; (2) to outline trends in evolution of anatomical features, as established or suggested by various authors; (3) to offer examples in which anatomy has aided in clarification of taxonomic problems; (4) to recommend particular methods for study of anatomical features and for illustration of results; (5) to emphasize both usefulness and limitations of anatomy in taxonomic studies, and thus to show that systematists cannot afford to overlook anatomical information; and (6) to imply that plant anatomists, whatever their interests, can benefit from knowledge of comparative data.

    Likewise, there are a number of purposes that this book is *not* intended to fulfill: (1) This book is to be considered supplementary to standard anatomy texts, and knowledge of principles and terms is presupposed. (2) Literature on any topic is a selection, and not exhaustive; consultation of literature cited herein and in the papers listed should quickly lead the reader to most of the papers on any given topic. (3) Although various papers have been selected here and analyzed, validity of their conclusions can be established only by reinvestigation. (4) Space does not allow discussion of microtechnical methods, although the method of preparation often significantly influences the findings of a comparative investigation. Finally, (5), this book is not intended to detract from compilations such as those listed below, but rather to enhance their usefulness.

The following books are indispensable for comparative anatomists:

1. Solereder, H., *Systematic anatomy of the dicotyledons* (trans. Boodle & Fritsch), 1908. Clarendon Press, Oxford.
2. Solereder, H., and J. Meyer, *Systematische Anatomie der Monokotyledonen,* Vol. I, 1928; III, 1929; IV, 1930; VI, 1933. Gebrüder Borntraeger, Berlin.
3. Metcalfe, C. R., and L. Chalk, *Anatomy of the dicotyledons,* 1950. Clarendon Press, Oxford.
4. Metcalfe, C. R., *Anatomy of the monocotyledons,* Clarendon Press, Oxford. Vol. I, *Gramineae,* 1960; Vol. II, *Palmae,* 1961; other families to appear in two subsequent volumes; contributions by various authors.
5. Linsbauer, K., ed., *Handbuch der Pflanzenanatomie. Gebrüder* Borntraeger, Berlin. (Monographs in the series appeared at various times, and are cited in the chapters of this book under the names of individual authors.)

Attention should also be called to Metcalfe and Chalk's forthcoming supplement to *Anatomy of the dicotyledons,* and to the parts, which have already begun to appear, of the second edition of the *Handbuch der Pflanzenanatomie.*

A number of figures have been redrawn from papers by various authors. For permission to use these, I wish to extend my appreciation expressly to those authors (cited beneath redrawn figures), and to editors of journals in which these papers appeared. Several figures are copied by permission of the Clarendon Press. Drs. Adriance S. Foster, William L. Stern, and Richard A. Howard deserve special thanks for reading Chapters 2, 4, and 6, respectively, and for offering helpful suggestions.

S. C.

Claremont, California
April, 1961

contents

COMPARATIVE
PLANT ANATOMY

chapter one

The Ethics
of Comparison

Comparative anatomical studies of angiosperms have achieved a remarkable record within the past century, and one may safely say that few systematic studies would fail to benefit from incorporation of anatomical data. Comparative anatomy has proved to be useful in some of the most difficult taxonomic studies. The reader may be surprised to know that many papers attest to the value of anatomical methods in analyzing natural hybridization (Barua and Wight, 1958; Cannon, 1909; Cousins, 1933; MacFarlane, 1892; Pryor, Chattaway, and Kloot, 1956; Rollins, 1944; Russell, 1919). Among the more spectacular achievements of comparative anatomy are the solutions of cases involving "misplaced," "anomalous," or "isolated" genera. Many instances of this sort could be cited, as even a casual perusal of Metcalfe and Chalk (1950) would indicate.

Before reviewing the remarkable ways in which plant anatomy serves studies of evolution and taxonomy, sources of variability must be considered.

PROBLEMS OF THE VARIATION PATTERN

Variations related to environment or location within a plant

The beginning student learns about variability of this nature in the classic concept of "sun" leaves and "shade" leaves. The changes in mesophyll structure as a result of shading are familiar,

1

but other differences among leaves with relation to ecological factors may not be. Zalenski (1902), who pioneered in this field, explored sources of variability in venation. Areole size, for example, may be much larger near the base of a plant, decreasing to a minimum in the most illuminated portions (Carlquist, 1957).

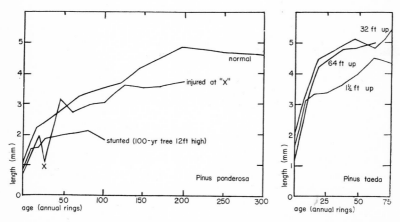

Fig. 1-1. Graphs showing effect of various factors on length of tracheids in *Pinus*. Graph at left shows that stunting by adverse conditions or injury results in markedly shorter tracheids. Graph at right shows that tracheids are markedly shorter near base of tree than farther up. (Redrawn from Bailey and Tupper, 1918.)

Cell size may vary from one part of a plant to another. Tenopyr (1918) has shown such variability in leaves, and Solereder (1908) claims that cells of adaxial leaf epidermis tend to be larger than those of the abaxial surface. Even a portion of the plant considered to be rather conservative anatomically—secondary xylem—may show rather startling modifiability. Even supposing that xylem from an old trunk is studied, the source of a sample is important. The graph at the right in Fig. 1-1 illustrates that tracheids near the base of a tree are much shorter than those higher in a tree for wood of comparable ages. The graph at the left shows effects of ecological conditions that may be termed "stunting." Bailey and Tupper (1918) report that longer tracheary elements characterize trees in closed stands, shorter ones occur in trees growing in isolated positions. The effect of injury (Fig. 1-1, left) is also apparently a long-lasting one. Morrison (1953) has shown that in a given species, significant differences in vessel diameter, vessel-element length, fiber length, fiber-wall thickness, ray height, ray width, and vessel concentration may be related to how a root is formed—that is, at the surface or beneath the surface of the soil. Similarly, Ingle and Dadswell (1953)

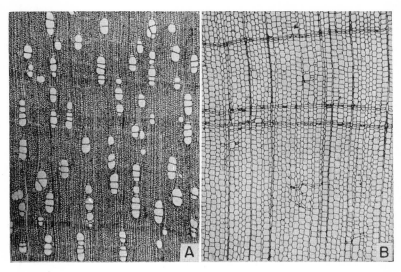

Fig. 1-2. Comparison of transections of secondary xylem of stem (A) and root (B) in *Alstonia spathulata* (Apocynaceae). Note the infrequency of vessels, wide diameter of fibers, and more prominent growth rings in root as compared with stem. Both ×17. (From Ingle and Dadswell, 1953.)

point out a striking instance of difference between stem and root (Fig. 1-2) in respect to xylem features, both qualitative and quantitative. Another type of variation is represented in Fig. 1-3. In these sections, entirely different ray shapes and ray cell types characterize areas (of the same age) from the same stem.

Variability can easily be so great as to negate the use of features for taxonomic purposes unless an anatomist can discriminate between heritable characteristics and those that are environmentally modifiable. Unfortunately, the fact that some features may be influenced by environment or location within a plant has led to skepticism as to the value for study of anatomical features at large.

Such skepticism is adequately answered by Metcalfe (1959), but the comparative anatomist must always be able to defend himself against such charges by understanding the modifiability of his material.

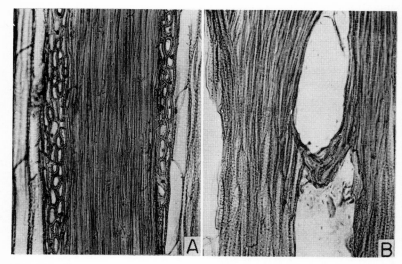

Fig. 1-3. Comparison of tangential sections of wood from two parts, both of the same age, of a single stem of *Tetradymia glabrata* (Compositae). In B, rays are wider and consist of thin-walled cells (collapsed), whereas in A, rays are narrow and are composed of thick-walled, lignified cells. Both ×128.

Variations related to maturation or age of tissues

"Maturity" may refer either to a cell or group of cells, or to the parts of a plant. Thus, maturity in a leaf is quite different from maturity in secondary xylem, which may exhibit an "adult" pattern of cell types and sizes only after many years of cambial activity.

When is the venation pattern of a cotyledon "mature"? Probably not at germination; the most valid ideas concerning cotyledon venation are probably best derived from a cotyledon just before cessation of function. Likewise, vascularization patterns of flowers are complete in some portions of the flower prior to anthesis, but other parts of the vascular system, present as procambium in a flower, may not mature until growth of the fruit is completed.

Thus, stages selected for studies in floral anatomy may be, to some extent, arbitrary.

Development of concepts of a mature pattern in secondary xylem has been slow and difficult. Such concepts were advanced by Bailey and Tupper (1918), who showed that length of tracheary elements in a plant varies quite appreciably with location (in respect to age) within a stem. For example, the graphs of Figs. 1-1 and 4-1, indicate that a stable pattern of tracheary length may not be reached for many years. Stern and Greene (1958), and the papers they cite, suggest that qualitative characteristics may mature much earlier than quantitative ones. Similarly, the studies of Sebastine (1955) should be familiar to any student contemplating statistical analysis of wood characteristics. These papers reveal that qualitative features, when compared from species to species, are often reliable, but unless source of wood sample is known precisely and unless variability of quantitative features in a wood is understood, statistical treatment is genuinely meaningless. It may seem shocking to say that much—if not most—of the anatomical work in which average, standard deviation, and other figures are calculated lacks statistical significance because of great variation (usually unexplored), but the statement is probably true. Secondary xylem offers some of the few features in plant anatomy to which quantitative measures seem to be applicable, but the anatomist must not enter the race with the "exact sciences" toward precision of quantitative measures—if he can enter that race at all—unless he knows the limitations of his material.

Interestingly, certain woods may show slow maturation in qualitative, as well as quantitative, features. In Fig. 1-4, sections representing different ages within a single sample exhibit ray types that would fall into different categories in the ray-type classification most widely used.

Variations among individuals

Other variations found within populations may be ascribed to (1) varied morphogenetic expression of a single genotype or (2) infraspecific genetic variability. Subspecific anatomical criteria do occur, but are rather rare. Whatever the cause of variation, the range, as well as the "typical" condition, should be described. The idea of presenting such a range in an anatomical feature—which may be termed a "complement" or a "population"—is emphasized throughout this book.

Examples of such techniques may be found in the work of Foster (1944), who determines a variety of petiolar sclereids in *Camellia japonica,* and Rodriguez (1957), who offers an interesting

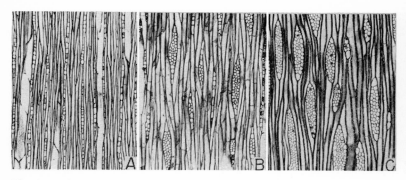

Fig. 1-4. Comparison of tangential sections of secondary xylem from a single stem of *Bursera simaruba* (Burseraceae) to show change in ray types and fiber diameter with age. A is a section from near the pith; B, a section 4 cm from the pith; C, a section 9 cm from the pith. Compare with ray types in Fig. 4-11. All ×50. (From Barghoorn, 1941.)

method of representing variation within a species in perforation-plate morphology. Kasapligil (1951) illustrates a number of outer stamens, inner stamens, and ovaries for *Umbellularia californica* in order to demonstrate variability in these organs.

PROBLEMS IN PHYLOGENETIC INTERPRETATION

As Eames (1957) says, "the study of morphology is, above all, the study of evolutionary modification of form. . . . Morphology must go hand in hand with other fields in the establishment of phylogenetic relationship." Fortunately, many plant anatomists, following the advice repeatedly offered by I. W. Bailey, have based phylogenetic hypotheses upon features that, when properly assessed, show unidirectional trends of evolution, such as those related to progressive shortening of tracheary elements. Other features, however, strongly admit of phylogenetic interpretation. For example, the primitiveness of the two-trace unilacunar node and the monosulcate pollen grain in angiosperms is finding wide acceptance. One must remember, however, that most anatomical characteristics are, in all likelihood, reversible to some extent. Moreover, even in unidirectional trends, interpretation of small differences in a particular taxonomic group is unwarranted (Bailey, 1957).

One must remember that only characteristics, not plants or species, are primitive or advanced. If this principle is kept in mind, phylogenetic hypotheses are likely to be conceived with greater accuracy. As Bailey (1957) notes, various features may show little or great deviation in their synchronization during evolution of a particular phylad. Thus, one can question Boureau's (1957) claim that the presence of primitive xylem in various angiosperm groups argues for polyphylesis. Bailey has also noted (1949, 1951) that failure to take into account all features of a plant, and weigh them according to their individual worth, can lead to flawed or incorrect conclusions. Indeed, if a plant is viewed as an evolutionary vehicle in which various features, primitive and advanced, are contained, the phyletic advancement of any species may be viewed as a compromise among the totality of its characteristics. Anatomical features may well reinforce interpretations based upon gross morphology (Metcalfe, 1954), but again they may not. Bailey (1957) has emphasized that anatomical information may be particularly valuable in negating certain phylogenetic hypotheses. Metcalfe (1954) shows how such data have aided in negating close relationship between Alismataceae and Ranunculaceae, and in severely limiting the likelihood of Corner's "Durian Theory." As the following chapters illustrate, anatomy can aid in formulation of positive conclusions also. Anatomical data must take their rightful place, not more or less, in phylogenetic studies.

REFERENCES

BAILEY, I. W., 1949. "Origin of the angiosperms: need for a broadened outlook," *Jour. Arnold Arb.*, **30:** 64–70.

——, 1951. "The use and abuse of anatomical data in the study of phylogeny and classification," *Phytomorphology*, **1:** 67–69.

——, 1957. "The potentialities and limitations of wood anatomy in the study of phylogeny and classification of angiosperms," *Jour. Arnold Arb.*, **38:** 243–254.

8 · THE ETHICS OF COMPARISON

BAILEY, I. W., and TUPPER, W. W., 1918. "Size variations in tracheary cells. I. A comparison between the secondary xylems of vascular cryptogams, gymnosperms, and angiosperms," *Proc. Amer. Acad. Arts and Sci.*, **54**: 149–204.

BARGHOORN, E. S., 1941. "The ontogenetic development and phylogenetic specialization of rays in the xylem of dicotyledons. II. Modification of the multiseriate and uniseriate rays," *Amer. Jour. Bot.*, **28**: 273–282.

BARUA, P. K., and WIGHT, W., 1958. "Leaf sclereids in the taxonomy of thea camellias. I. Wilson's and related camellias," *Phytomorphology*, **8**: 257–264.

BOUREAU, E., 1957. *Anatomie végétale*, Vol. 3, Paris: Presses Universitaires de France.

CANNON, W. A., 1909. "Studies in heredity as illustrated by the trichomes of species and hybrids of *Juglans, Oenothera, Papaver*, and *Solanum*." *Carnegie Inst. Wash. Publ.*, **117**: 1–67.

CARLQUIST, S., 1957. "The genus *Fitchia* (Compositae),". *Univ. Calif. Publ. Bot.* **29**: 1–144.

COUSINS, S. M., 1933. "The comparative anatomy of the stem of *Betula pumila, Betula lenta,* and the hybrid *Betula jackii*," *Jour. Arnold Arb.*, **14**: 351–355.

EAMES, A. J., 1957. "Some aspects of progress in plant morphology during the past fifty years," *Amer. Jour. Bot.*, **44**: 100–104.

FOSTER, A. S., 1944. "Structure and development of sclereids in the petiole of *Camellia japonica* L.," *Bull. Torrey Bot. Club*, **71**: 302–326.

INGLE, H. D., and DADSWELL, H. E., 1953. "The anatomy of the timbers of the south-west Pacific area. II. Apocynaceae and Annonaceae," *Austral. Jour. Bot.*, **1**: 1–26.

KASAPLIGIL, B., 1951. "Morphological and ontogenetic studies of *Umbellularia californica* Nutt. and *Laurus nobilis* L.," *Univ. Calif. Publ. Bot.*, **25**: 115–240.

MACFARLANE, J. M., 1892. "A comparison of the minute structure of plant hybrids with that of their parents and its bearing on biological problems," *Trans. Roy. Soc. Edinb.*, **37**: 203–286.

METCALFE, C. R., 1954. "An anatomist's view of angiosperm classification," *Kew Bull.*, **3**: 427–440.

————, 1959. "A vista in plant anatomy," in *Vistas in botany*, London: Pergamon Press.

————, and CHALK, L., 1950. *Anatomy of the dicotyledons*, Oxford: Clarendon Press.

MORRISON, T. M., 1953. "Comparative histology of secondary xylem in buried and exposed roots of dicotyledonous trees," *Phytomorphology*, **3**: 427–430.

PRYOR, L. D., CHATTAWAY, M. MARGARET, and KLOOT, N. H., 1956. "The inheritance of wood and bark characters in *Eucalyptus*," *Austral. Jour. Bot.*, **4**: 216–239.

RODRIGUEZ, R. L., 1957. "Systematic anatomical studies on *Myrrhidendron* and other woody Umbellales," *Univ. Calif. Publ. Bot.*, **29**: 145–318.

ROLLINS, R. C., 1944. "Evidence for natural hybridity between guayule (*Parthenium argentatum*) and mariola (*Parthenium incanum*)," *Amer. Jour. Bot.*, **31**: 93–99.

RUSSELL, A. M., 1919. "The macroscopic and microscopic structure of some hybrid sarracenias compared with that of their parents," *Contrib. Bot. Lab. Univ. Penn.,* **5:** 1–41.

SEBASTINE, K. M., 1955. "Studies on the variations in the structure and size of rays in secondary wood," *Jour. Indian Bot. Soc.,* **34:** 299–306.

SOLEREDER, H., 1908. *Systematic anatomy of the dicotyledons* (trans. by Boodle and Fritsch), Oxford: Clarendon Press.

STERN, W. L., and GREENE, S., 1958. "Some aspects of variation in wood," *Trop. Woods,* **108:** 65–71.

TENOPYR, LILLIAN A., 1918. "On the constance of cell shape in leaves of varying shape," *Bull. Torrey Bot. Club,* **45:** 51–76.

ZALENSKI, V., 1902. "Über die Ausbildung der Nervation bei verscheidenen Pflanzen," *Ber. Deutsch. Bot. Ges.,* **20:** 433–440.

chapter two
Cellular Components, Idioblasts, Laticifers, and Secretory Structures

Phenomena treated in this chapter may occur in more than one organ of a plant, and each of the topics is related to others in some measure. For example, discussion of sclereids is related to crystals in some cases because some sclereids (for example, "cristarque cells") characteristically contain crystals. Cellular and intercellular structures may be allied, as is the case in mucilage cells and mucilage cavities. To a considerable extent, definition of items considered in this chapter are related to function or chemical nature (Foster, 1956; Sperlich, 1939). Unhappily, comparative chemical data are badly needed to supplement our knowledge of morphology of these structures.

DISTINCTIVE CELL-WALL CHARACTERISTICS

Pectic warts

Intercellular extrusion of pectic materials, often termed pectic warts, may be of systematic value. For example, pectic warts in Compositae (Carlquist, 1956) are different from those in monocots.

10

For a review of this topic, see Kisser (1928). Amorphous intercellular pectic secretions proved of systematic value in distinguishing *Wilkesia* from *Argyroxiphium* (Carlquist, 1957).

Sclereids

Sclereids may serve in various ways to aid taxonomy. Elongate sclereids in the epidermis of ovules of Cynareae (Compositae) aid in defining that tribe. Idioblastic sclereids, especially those in leaves, offer many possibilities. Foster (1946) finds correlation between four distinctive sclereid types and the four sections of *Mouriri* (Fig. 2-1). Barua and Wight (1958) analyze *Camellia* species on the basis of variations in morphology and location of sclereids. Rao (1957) has demonstrated the taxonomic utility of these cells in *Memecylon,* and Tomlinson's (1959b) work demonstrates their systematic value in palms. Sclereids terminal on vein endings in leaves occur in a number of unrelated families (Foster, 1956), but they may be useful systematically. Foster (1955) has found that some species of *Boronella* have terminal sclereids nearly exclusively, whereas other species may have both terminal and diffuse foliar sclereids.

Storage tracheids and "spiral cells"

Under such terms as these, cells with a variety of histological characteristics have been described. Such cells as the ovoid idioblasts of *Pogonophora* (Foster, 1956), the circular-pitted storage tracheids of *Capparis galeata* (Metcalfe and Chalk, 1950), the spiral cells of *Glishrocolla* and *Nepenthes* (Metcalfe and Chalk, 1950), or the reticulate-walled tracheids in *Spiranthes spiralis* (Pirwitz, 1931) are worthy of further exploration to determine if these features occur in relatives of these plants. For a review of these phenomena, see Pirwitz (1931).

CELLULAR CONTENTS

Plastids

Plastids in angiosperms are of systematic value not because of variations in their shape, but rather because the distribution patterns of cells containing chloroplasts and other types of plastids are of interest. For a review of literature on these topics, see Schürhoff (1924).

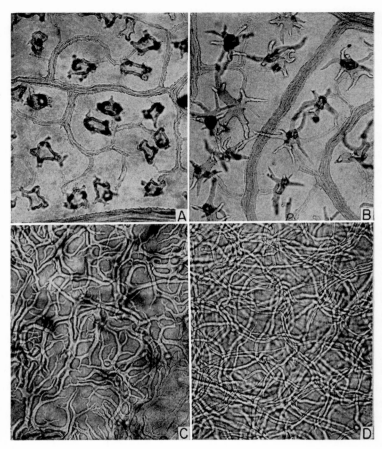

Fig. 2-1. Portions of cleared leaves, to show four types of sclereids in species of *Mouriri* (Melastomaceae). A, Type I, *M. emarginata;* B, Type II, *M. princeps;* C, Type III, *M. Gardneri;* D, Type IV, *M. apiranga.* All ×75. (From Foster, 1946.)

Starches

Although many authors have investigated the potentialities of starch grains for systematic use, the monumental monograph of Reichert (1913) offers much more new data than does the work of other authors, besides summarizing earlier descriptive work and classification systems. Many books illustrate variety in starch-grain shape, but few accounts other than Reichert's show that generic, specific, and even varietal criteria may be derived from careful study of particular groups. The accompanying drawings (Fig. 2-2)

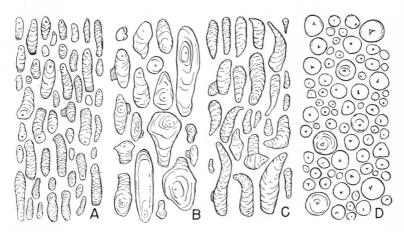

Fig. 2-2. Populations of starch grains from cortex of four species of Araceae, showing differentiation corresponding to the taxonomic system. A, *Dieffenbachia seguine* var. *nobilis;* B, *D. seguine* var. *maculata;* C, *D. illustris;* D, *Arisaema triphyllum.* All ×300. (Drawn from Reichert, 1913.)

have been selected to show these principles. A feature of Reichert's work that deserves special mention is his use of graphic representations of chemical reactions of these starches as a comparative tool.

Proteins

Comparisons of solid protein depositions are summarized by Wieler (1943). Proteid bodies characterize some Cactaceae (Metcalfe and Chalk, 1950). Guérin (1923) has figured "albuminoids" from laticifers (Fig. 2-3) that seemingly distinguish three species of *Laportea.* A study by Steckbeck (1919) reveals that in Oxalidaceae and Leguminosae that possess leaf or leaflet movement, more prominent proteinaceous bodies are present in species with greater sensitivity in leaf movement, and that this increase in size of proteinaceous bodies has paralleled evolutionary increase in sensitivity.

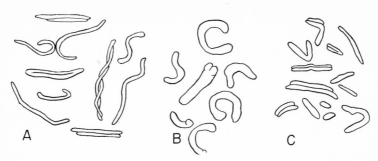

Fig. 2-3. Comparison of populations of albuminoid bodies from laticifers of three species of *Laportea* (Urticaceae). A, *L. moroides;* B, *L. Schomburghii versicolor;* C, *L. platycarpha.* A, ×420; B, C, ×280. (Redrawn from Guérin, 1923.)

Pigments

As the interesting systematic study of Reznik (1957) suggests, comparative study of plant pigments is probably best conducted within the various disciplines of chemistry. Nevertheless, Thaler *et al.* (1959) offer cases of anthocyanin idioblasts that suggest the value of exploration in this field.

Crystals; other mineral deposition

Literature on these topics has been summarized in the *Handbuch der Pflanzenanatomie* by Frey (1929) and Netolitzky (1929). Many subsequent contributions deserve mention.

Silica. Silica bodies and silica cells have been found to be of systematic value in a number of groups, such as epidermis of grasses (see Fig. 3-1, p. 26). As the four species illustrated in Fig. 2-4 suggest, variations in size, number per cell, and "spininess" of silica bodies may be systematic characteristics in Rapateaceae. Tomlinson (1956) has called attention to epidermal cells, each containing a large silica body, that occur in longitudinal patterns paralleling veins; these cells are termed "stegmata." Such cells obviously have great systematic use, as Tomlinson shows in Zingiberaceae (1956), Musaceae (1959a), and Palmae (1959b), and can be used both at the generic and specific level. Silica bodies have been found to be useful in dicots, as in the chrysobalanoid Rosaceae. Of a different nature are the grouped silicified cells of Aristolochiaceae and Loranthaceae that characterize certain genera in those families (Metcalfe and Chalk, 1950).

Gypsum. Gypsum crystals, rare in angiosperms at large, are characteristic of certain Capparidaceae (Solereder, 1908).

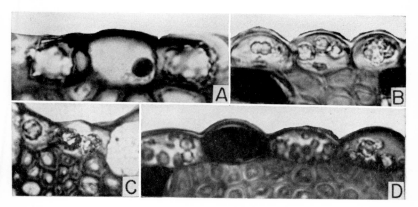

Fig. 2-4. Comparison of silica bodies in epidermal cells of leaves of Rapateaceae. A, *Rapatea paludosa;* B, *Guacamaya superba;* C, *Rapatea spruceana;* D, *Schoenocephalum coriaceum.* Note markedly larger bodies in A, markedly smaller in D. All ×725.

Calcium oxalate. Oxalate crystals account for most crystalline or mineral objects occurring in plants. Because a single plant may contain a small or great variety of oxalate crystals, the writer recommends that this variety be described and figured, and the occurrence of the various types throughout the plant determined. Such a "crystal complement" is illustrated here for *Hosta subcordata* (Fig. 2-5). This descriptive approach has been used by Heintzelman and Howard (1948) in Icacinaceae. Use of polarized light (see, for example, Pobeguin, 1943) is recommended for location of crystals that may otherwise be difficult to observe. For a useful index, with classification of crystal types, see Solereder (1908), Metcalfe and Chalk (1950), and the literature dealing with crystals in wood, cited in Chapter 4.

Abundance of raphides in Onagraceae (in contrast with their absence in supposedly related families) or absence of raphides in Hydrocharitaceae (in contrast to their abundance in monocots) were early noted by Gulliver (1864) as significant in defining larger groupings of angiosperms. Within more limited taxonomic vistas, crystals in the ovary wall of Cichorieae have proved significant to Stebbins (1940); Rothert (1900) has used styloids in comparing species of *Eichornia*. Perhaps the most unusual and intriguing instance of use of crystals has been offered by Chartschenko (1932). In this paper,

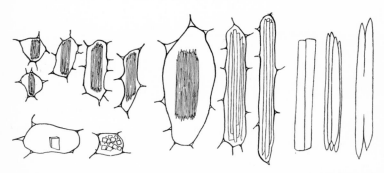

Fig. 2-5. The "crystal complement" of *Hosta* (*Funckia*) *subcordata* (Liliaceae). At left, above, raphides; below, cuboidal crystals; at right, prismatic crystals and styloids. ×220. (Redrawn from Pobeguin, 1943.)

the species and species groups in *Allium* are found to be definable by means of oxalate crystals and, in addition, a phylogenetic tree of crystal complements (Fig. 2-6) is offered. This tree suggests that the potential of crystals in phylogenetic studies deserves further exploration.

Calcium carbonate. This compound may occur diffusely in cells, as in Cucurbitaceae, or as a component of a cystolith. Cystoliths may not incorporate carbonate, as in the chalk-free cystoliths of Acanthaceae (Linsbauer, 1921), resin cystoliths of Begoniaceae, or slime cystoliths of Urticaceae (Linsbauer, 1921). In at least some cases, cystoliths aid in linking related families (for example, Moraceae, Urticaceae, and Cannabinaceae). The three examples illustrated in Fig. 2-7 suggest the value of cystoliths within Urticaceae. For information on epidermal cystoliths, see Linsbauer (1930); summaries of other aspects of cystoliths are offered by Berg (1932) and Beyrich (1944).

Crystalliferous sclereids. Distinctive cell types may possess crystals embedded in walls or secreted within the lumen. An example of the former condition is shown in Fig. 9-1B (p. 117), a crystalliferous trichome of the sort possessed by many Rubiaceae. Sclereids in

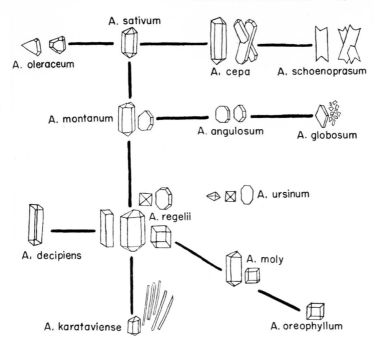

Fig. 2-6. A "crystal phylogeny" for species of *Allium* (Liliaceae). According to this phylogeny, a decreasing diversity of crystal types would characterize the more derived species. (Redrawn from Chartschenko, 1932.)

Nymphaeaceae are notable for their crystal-studded walls. In Loranthaceae, a solitary crystal is often present in the lumina of sclereids, a condition also seen in *Cereus* (Cactaceae). Perhaps the most significant instances of this sort, however, are the cristarque cells—crystal-bearing sclereids having eccentrically thickened walls —that serve to link Quiinaceae, Ochnaceae, and Scytopetalaceae. These phenomena are similar to stegmata of certain palms (Tomlinson, 1959b).

Tanniniferous cells

Contents of these cells are best determined from microchemical reactions (see Oberstein, 1913), although most plant anatomists trust more pragmatic staining reactions. Microchemical work may reveal distinctive types of tanniniferous cells. Thus, Lemesle (1937), study-

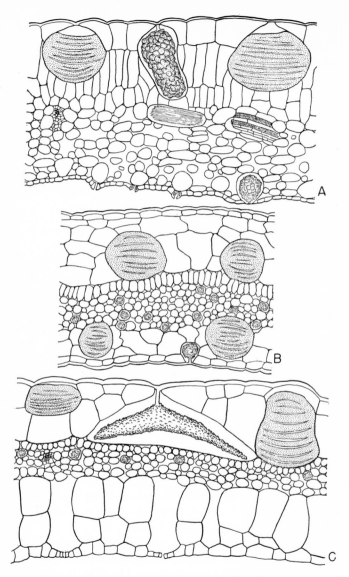

Fig. 2-7. Comparison of leaf structure and foliar idioblasts in Urticaceae. Note mucilage cells (stippled), druses, and cystoliths. A, *Laportea longifolia;* B, *Procris pedunculata;* C, *Elatostema repens.* A, ×190; B, C, ×100. (Redrawn from Guérin, 1923.)

ing chemistry and histology of such cells in *Eupomatia,* recognizes three types. Such cells often occur idioblastically (Fig. 7-1C, D, p. 96); presence of tanniniferous cells in root cortex of Rapateaceae versus their absence in the related family Xyridaceae seems a systematic criterion.

Fats and oils

Possibly the best single characteristic that may be used to unite the heterogeneous order Ranales is the presence of ethereal oil cells that occur idioblastically. For a summary of studies on ethereal oil cells, see Lehmann (1925). Fat bodies occur in *Ilex paraguayensis* (Metcalfe and Chalk, 1950) and may be more widespread systematically.

LATICIFERS

No two summaries of the systematic distribution of laticifers agree, because our ideas on what constitutes a laticifer and what latex is are very vague, and the problem is insoluble in terms of present literature. In study of laticifers, developmental knowledge is important in defining laticifer type; unhappily, such data are not always available. For recent classifications of laticifer types, see Schaffstein (1932) and Boureau (1954). An excellent summary of morphology and types of laticifers is offered by Esau (1953).

Laticifers do serve for systematic comparisons. The families Moraceae, Urticaceae, and Cannabinaceae are allied on the basis of a particular laticifer type, just as Apocynaceae and Asclepiadaceae exhibit similaries in laticifer distribution and structure. Within particular families laticifers may be significant. Metcalfe and Chalk (1950) analyze systematic occurrence of four types of laticifers in Euphorbiaceae. Likewise, Compositae possess both laticiferous cells and articulated anastamosing laticifers, and systematic and organographic distribution of these is basic in development of ideas on phylogeny and relationships within the family (Col, 1904). The work of De Bary (1884) in Araceae suggests that a phylogenetic explanation may underlie the diverse laticifer types in that family and their systematic occurrence. Warsow (1903), working with *Acer,* provides another instance of taxonomic use of laticifers.

The lists of Solereder (1908) and Metcalfe and Chalk (1950) of families with laticiferous cells and canals are greater than one might expect. This must stem—at least in part—from the unknown nature of latex in various groups. Thus, laticifers of Fumariaceae (Heinricher, 1887, 1891) may be similar to those of Papaveraceae, and the systematic relationship of these families to Cruciferae tempts one to find a morphological analog in the "protein-bearing" cells, laticiferlike in shape, in Cruciferae (Heinricher, 1884). In any case,

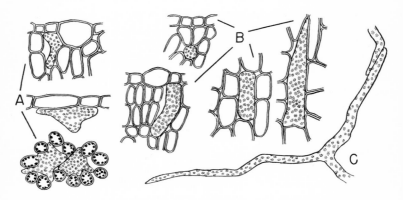

Fig. 2-8. Comparison of protein-bearing idioblasts in Cruciferae. Planes of section from which drawings are derived are various, but distinctions in shapes and sizes among the species seem evident. Stippled areas = protein granules; cells, below in A, with large dots = chlorenchyma. A, *Moricandia arvensis;* B, *Diplotaxis tenuifolia;* C, *Sinapis alba.* All ×150. (Redrawn from Heinricher, 1884.)

these cells show variations corresponding to taxonomic units, as shown in Fig. 2-8. Further studies on laticiferous structures in a number of families (for example, Alismataceae, Cactaceae) need documentation with chemical tests. An example of problematic idioblasts germane to this problem is found in *Tetracentron* (Bailey and Nast, 1945; Foster, 1956).

MUCILAGE CELLS, CANALS, AND CAVITIES

Mucilage occurs both in cells, as in Cactaceae (Lauterbach, 1889), and in canals, as in Piperaceae (Van Tieghem, 1908). Boundaries between these two types may be insignificant, because gelatinization of walls between adjacent mucilage cells can result in a lysigenous canal. Size, distribution, etc., of mucilage cells and canals may be useful in systematics, as the work of Guérin (1923) suggests (see Fig. 2-8).

SECRETORY CANALS AND CAVITIES

Lysigenous mucilage canals mentioned above are but one of many types of canals that occur. In fact, schizogenous canals and cavities probably predominate, and the listings of various types of canals and cavities by Metcalfe and Chalk (1950) probably consist mostly of such structures—usually with contents vaguely defined. Other than mode of origin and contents, useful features of canals and cavities may include distribution within the plant, size, and number of layers ("epithelial cells") lining the canal. Organographic distribution has been most significant in investigations, and in all likelihood secretory canals do, in some groups, show trends of decrease or increase in occurrence. Thus, Col (1904) and Jeffrey (1917) hypothesize abundance of secretory canals primitively in Compositae, with differential rates of loss in different organs for various species. Stern (1952) utilizes presence of secretory canals to aid in allying Julianiaceae with Anacardiaceae and Burseraceae.

For older contributions on secretory canals, containing much comparative data, see Van Tieghem (1872, 1885) and Leblois (1887). Recent contributions in particular groups include those of Mayberry (1936) and Moens (1955).

REFERENCES

BAILEY, I. W., and NAST, CHARLOTTE G., 1945. "Morphology and relationships of *Trochodendron* and *Tetracentron*. I. Stem, root, and leaf," *Jour. Arnold Arb.*, **26:** 243–254.

BARUA, P. K., and WIGHT, W., 1958. "Leaf sclereids in the taxonomy of thea camellias. I. Wilson's and related camellias," *Phytomorphology*, **8:** 257–264.

BERG, A., 1932. "Untersuchungen über die Entwickelungsbedingungen der Zystolithen," *Beih. Bot. Centr.*, **49**(1): 239–259.

BEYRICH, H., 1944. "Neue Beiträge zur Entwicklungsgeschichte und Morphologie der Zystolithen," *Protoplasma*, **38:** 287–313.

BOUREAU, E., 1954. *Anatomie végétale*, Vol. I, Paris: Presses Universitaires de France.

CARLQUIST, S., 1956. "On the occurrence of intercellular pectic warts in Compositae," *Amer. Jour. Bot.*, **43:** 425–429.

CARLQUIST, S., 1957. "Leaf anatomy and ontogeny in *Argyroxiphium* and *Wilkesia* (Compositae)," *Amer. Jour. Bot.*, **44**: 696–705.

CHARTSCHENKO, W., 1932. "Verschiedene Typen des mechanischen Gewebes und des kristallinischen Ausbildungen als systematische Merkmale der Gattung *Allium*," *Beih. Bot. Centr.*, **50**(2): 183–206.

COL, M. S., 1904. "Sur l'appareil sécréteur interne des composées," *Jour. de Bot.*, **18**: 153–175.

DE BARY, A., 1884. *Comparative anatomy of the vegetative organs of the phanerogams and ferns*, Oxford: Clarendon Press.

ESAU, KATHERINE, 1953. *Plant anatomy*, New York: John Wiley.

FOSTER, A. S., 1946. "Comparative morphology of the foliar sclereids in the genus *Mouriria* Aubl.," *Jour. Arnold Arb.*, **27**: 253–271.

———, 1955. "Comparative morphology of the foliar sclereids in *Boronella* Baill.," *Jour. Arnold Arb.*, **36**: 189–198.

———, 1956. "Plant idioblasts: remarkable examples of cell specialization," *Protoplasma*, **46**: 184–193.

FREY, A., 1929. "Calcium-Monohydrat und Trihydrat," in K. Linsbauer, ed., *Handbuch der Pflanzenanatomie*, 3(1a): 81–118.

GUÉRIN, P., 1923. "Les urticées: cellules à mucilage, laticifères, et canaux sécréteurs," *Bull. Soc. Bot. France*, **70**: 125–136 *et seq.*

GULLIVER, G., 1864. "On Onagraceae and Hydrocharidaceae as elucidating the value of raphides as natural characters," *Jour. of Bot.*, **2**: 68–70.

HEINRICHER, E., 1884. "Über Eiweissstoffe führende Idioblasten bei einigen Cruciferen," *Ber. Deutsch. Bot. Ges.*, **2**: 463–466.

———, 1887. "Vorläufige Mittheilung über die Schlauchzellen der Fumariaceen," *Ber. Deutsch. Bot. Ges.*, **5**: 233–239.

———, 1891. "Nochmals über die Schlauchzellen der Fumariaceen," *Ber. Deutsch. Bot. Ges.*, **9**: 184–187.

HEINTZELMAN, C. E., Jr., and HOWARD, R. A., 1948. "The comparative morphology of the Icacinaceae. V. The pubescence and crystals," *Amer. Jour. Bot.*, **35**: 42–52.

JEFFREY, E. C., 1917. *The anatomy of woody plants*, Chicago: University of Chicago Press.

KISSER, J., 1928. "Untersuchungen über das Vorkommen und die Verbreitung von Pektinwarzen," *Jahrb. Wiss. Bot.*, **68**: 206–232.

LAUTERBACH, C., 1889. "Untersuchungen über Bau und Entwicklung der Sekretbehälter bei den Cacteen," *Bot. Centr.*, **37**: 257–264 *et seq.*

LEHMANN, C., 1925. "Studien über den Bau und die Entwicklungsgeschichte von Ölzellen," *Planta*, **1**: 343–373.

LEMESLE, R., 1937. "Étude microchimique des divers tannoïdes de l'*Eupomatia*, *Bull. Soc. Bot. France*, **84**: 535–538.

LINSBAUER, K., 1921. "Über die kalkfreien Cystolithen der Acanthaceen," *Ber. Deutsch. Bot. Ges.*, **39**: 41–49.

———, 1930. "Die Epidermis," in K. Linsbauer, ed., *Handbuch der Pflanzenanatomie*, **4**: vii + 277.

MAYBERRY, M. W., 1936. "Hydrocarbon secretions and internal secretory systems of the Carduaceae, Ambrosiaceae, and Cichoriaceae," *Bull. Univ. Kansas,* **37:** 71–112.

METCALFE, C. R., and CHALK, L., 1950. *Anatomy of the dicotyledons,* Oxford: Clarendon Press.

MOENS, P., 1955. "Les formations sécrétrices des copaliers congolais," *La Cellule,* **57:** 35–64.

NETOLITZKY, F., 1929. "Die Kieselkörper als Zellinhaltskörper," in K. Linsbauer, ed., *Handbuch der Pflanzenanatomie,* 3(1a): viii + 80, 108–118.

OBERSTEIN, O., 1913. "Über das Auftreten von Gerbstoffidioblasten bei den Mesembriathemen," *Beih. Bot. Centr.,* **31**(1): 388–393.

PIRWITZ, K., 1931. "Physiologische und anatomische Untersuchungen an Speichertracheiden und Velamen," *Planta,* **14:** 19–76.

POBEGUIN, T., 1943. "Les oxalates de calcium chez quelques angiospermes," *Ann. Sci. Nat. Bot.,* sér. 11, **4:** 1–95.

RAO, T. A., 1957. "Comparative morphology and ontogeny of foliar sclereids in seed plants. I. *Memecylon,*" *Phytomorphology,* **7:** 306–330.

REICHERT, E. T., 1913. "The differentiation and specificity of starches in relation to genera, species, etc.," *Carnegie Inst. Wash. Publ.,* **173:** xvi + 900.

REZNIK, H., 1957. "Die Pigmente der Centrospermen als systematisches Element. II. Untersuchungen über das ionophoretische Verhalten," *Planta,* **49:** 406–434.

ROTHERT, W., 1900. "Die Krystallzellen der Pontederiaceen," *Bot. Zeit.,* **58:** 75–106.

SCHAFFSTEIN, G., 1932. "Untersuchungen an ungegliederten Milchröhren," *Beih. Bot. Centr.,* **49**(1): 197–220.

SCHÜRHOFF, P., 1924. "Die Plastiden," in K. Linsbauer, ed., *Handbuch der Pflanzenanatomie,* 1: v + 224.

SOLEREDER, H., 1908. *Systematic anatomy of the dicotyledons* (trans. by Boodle and Fritsch), Oxford: Clarendon Press.

SPERLICH, A., 1939. "Das trophische Parenchym. B. Excretionsgewebe," in K. Linsbauer, ed., *Handbuch der Pflanzenanatomie,* 4.

STEBBINS, G. L., Jr., 1940. "Studies in the Cichorieae. *Dubyaea* and *Soroseris,* endemics of the sino-himalayan region," *Mem. Torrey Bot. Club,* **19**(3): 1–76.

STECKBECK, D. W., 1919. "The comparative histology and irritability of sensitive plants," *Contrib. Bot. Lab. Univ. Penn.,* **4:** 185–230.

STERN, W. L., 1952. "The comparative anatomy of the xylem and the phylogeny of the Julianiaceae," *Amer. Jour. Bot.,* **39:** 220–229.

THALER, IRMTRAUD, WEBER, F., and WIDDER, F., 1959. "Anthozyan-Idioblasten der Frucht von *Polygonatum verticillatum,*" *Öst. Bot. Zeit.,* **106:** 124–132.

TOMLINSON, P. B., 1956. "Studies in the systematic anatomy of the Zingiberaceae," *Jour. Linn. Soc. London,* **55:** 547–592.

———, 1959a. "An anatomical approach to the classification of the Musaceae," *Jour. Linn. Soc. London,* **55:** 779–809.

———, 1959b. "Structure and distribution of sclereids in the leaves of palms," *New Phyt.,* **58:** 253–266.

VAN TIEGHEM, P., 1872. "Mémoire sur les canaux sécréteurs des plantes," *Ann. Sci. Nat. Bot.,* sér. 5, **16:** 96–201.

VAN TIEGHEM, P., 1885. "Second mémoire sur les canaux sécréteurs des plantes," *Ann. Sci. Nat. Bot., sér.* 7, 1: 5–96.

———, 1908. "Sur les canaux à mucilage des piperées," *Ann. Sci. Nat. Bot., sér.* 9, 7: 117–127.

WARSOW, G., 1903. "Systematisch-anatomische Untersuchungen des Blattes bei der Gattung *Acer* mit besonderer Berücksichtigung der Milchsaftelemente," Jena: Gustav Fischer.

WIELER, A., 1943. "Der feinere Bau der Aleuronkörner und ihre Entstehung," *Protoplasma,* **38:** 21–63.

> > > > >

chapter three

Epidermis
and Trichomes

Although one can often conveniently distinguish the epidermis from its appendages, trichomes, there are advantages to considering these categories as a unit. Trichomes are not always easily demarcated from papillate epidermal cells. That at least some plant anatomists consider trichomes together with other features of the epidermis is indicated by the monograph of Prat (1932). The several categories below, however, may conveniently be used for discussing various aspects of the epidermis. Although most studies on epidermis refer chiefly to that on leaves, this is by no means always true, and investigations of foliar epidermis should be complemented, in systematic studies, with study of epidermis from other portions of the plant.

EPIDERMIS

The monumental compilation of Linsbauer (1930) can be considered an encyclopedia of information on epidermis written up to that time. Individual contributions, chiefly those since this monograph or those emphasizing a comparative approach, are discussed below.

Outer epidermis wall and its coverings

Thickness, wall characteristics, and nature of sculpturing on the wall as seen in surface view have often been used as taxonomic criteria. Unfortunately, investigators have not always made distinctions between the epidermal wall itself and the cuticular covering. Sculpturing may result from ridges in the epidermis wall itself, as in Xyridaceae (Carlquist, 1960), or it may be formed wholly from cuticle, resulting in cuticular relief. Examples of the latter phenomenon as a taxonomic criterion have been offered by Raciborsky (1895) and Munz and Laudermilk (1949). Cuticular relief is a particularly interesting characteristic of epidermis of floral organs.

The presence of waxy platelets, rods, etc. (Linsbauer, 1930) may offer features of systematic interest. Bailey and Nast (1944) have shown the occlusion of stomata by an unidentified substance, which also forms warts and ridges on the epidermal surface, to be a characteristic of Winteraceae. Comparative data on epidermal walls that form layers during protracted existence is offered by Damm (1901). There may be a close relation, as shown by Cooper (1922), between thickness of cuticle and ecological situation; interpretation in this regard should therefore be considered.

Cell shape, size, and histology

Differentiation of epidermis can occur with regard to the tissue as a whole. For example, sclerification of epidermal walls is a feature

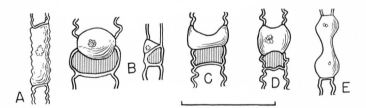

Fig. 3-1. Examples of suberized cells (vertical lines) and silica cells in epidermis of species of Gramineae. A, *Agropyron repens*, isolated cell from above a vein; B, *A. repens*, sectional view shown at right in addition to surface view at left; C, *Phyllostachys viridis glaucescens;* D, *Eragrostis pilosa;* E, *Digitaria sanguinale*. Bracket = 40 microns. (Redrawn from Prat, 1932.)

of taxonomic importance in Mutisieae (Carlquist, 1958a). Extremely long, narrow cells characterize epidermis of Stylidiaceae (Mildbraed, 1908). Distinctive patterns of "folding" in epidermal cell walls are discussed by Koehne (1884).

Idioblastic differentiation of epidermal cells (other than trichomes) offers many features. Prat (1932) has shown that in epidermis of certain Gramineae (Fig. 3-1), shape of silicified cells and suberized cells and specificity in relation of these to each other provide good taxonomic criteria. The occurrence of silica-bearing cells termed "stegmata" has been discussed in Chapter 2. Different types and

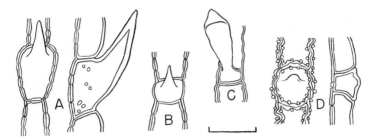

Fig. 3-2. Types of papillate epidermal cells (or short trichomes) in epidermis of Gramineae. A, *Secale cereale*, surface and side view of pointed cell; B, same, "hook" cell; C, *Lygeum spartum*, bicellular trichome from glume; D, *Triticum monococcum*, papillate cell, surface and side view. Bracket = 50 microns. (Redrawn from Prat, 1932.)

sizes of "light receptors" are specific criteria in the genus *Mesembryanthemum* (Kean, 1931). Bladderlike epidermal cells occur in families listed by Metcalfe and Chalk (1950), who also list, as does Solereder (1908), families with papillate epidermal cells. Whether papillate cells are considered trichomes or modified epidermal cells is unimportant, and transitions between them (see Fig. 3-4, upper right) may be found. Systematic importance of such cells is emphasized by Prat (1932) and types he figures are shown in Fig. 3-2.

Multiple epidermis

Continued periclinal division of epidermis is of taxonomic importance in certain families, such as Piperaceae.

Stomata

Much comparative data on angiosperm stomata have been accumulated, and some comparative studies, such as those of Warneke (1911) and Solereder (1908), may be cited. Smith (1935) has

shown that orientation of stomata—although influenced by underlying veins or shape of organ—may be a taxonomic characteristic,
as in Cactaceae.

Four main types of stomatal occurrence with relation to subsidiary cell number and arrangement have been recognized in dicots
by Solereder and by Metcalfe and Chalk. These are shown in Fig.
3-3A-D. Various modifications of these types (for example, Fig. 3-3F)

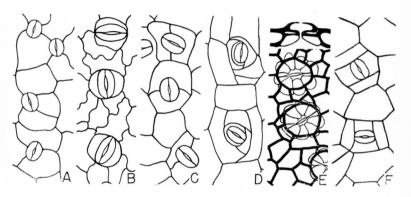

Fig. 3-3. Types of stomatal arrangements with relation to subsidiary cells in
angiosperms. A, *Eschscholzia californica*, showing a ranunculaceous (anomocytic)
type; B, *Calycanthus occidentalis*, rubiaceous (paracytic) type; C, *Ruellia graecizans*,
caryophyllaceous (diacytic) type; D, *Dudleya hassei*, cruciferous (anisocytic)
type; E, *Buxus balaerica*, showing a "special type" in which a circle of subsidiary
cells is cut off internal to the guard cells (sectional view above); F, *Rhoeo discolor*,
showing two pairs of subsidiary cells, a condition common in many monocots.

characterize certain groups. In addition, certain families have
"special types," such as that shown for Buxaceae (Fig. 3-3E). Characteristics of these sorts have been considered taxonomic or even
phylogenetic criteria in angiosperms (Francey, 1936; Porsch, 1905;
Rehfous, 1923), but considerable care must be used in interpretation.
Main types may show intergradation (even on a single leaf) and may
be attained in different ways. One might speculate that such a type
as the "cruciferous" represents a specialization over the "ranunculaceous" type. However, taxonomic and phylogenetic conclusions
should undoubtedly be framed within ontogenetic studies because
of the importance of developmental history.

Systematic studies

Perusal of complications such as Solereder and Meyer's *Systematische Anatomie der Monokotyledonen* demonstrates the very great importance of stomata and subsidiary cell features, particularly in monocots. Such a study as that of Prat (1932) suggests the potentialities of these investigations.

TRICHOMES

Of all anatomical features, trichomes are perhaps most often enlisted for systematic comparisons because of their variety, their almost universal presence in angiosperms, their ease of preparation and study, and the close relation of their variation patterns to the taxonomic system. The major compendia summarize a vast quantity of literature. In addition, special studies may be cited, such as Staudermann's (1924) study of monocot trichomes. Netolitzky (1932) based his survey largely on function, but offers much comparative data. For descriptive purposes, a classification such as that of Metcalfe and Chalk (1950) seems desirable, and the listings of families they—like Solereder (1908)—provide are very valuable. Although function, gross form, and cellular constitution are generally involved in terminology, grouping of terms (for example, uniseriate stalked glandular) provides relatively precise definitions. As emphasized by Heintzelman and Howard (1948), the taxonomist's traditional terminology for epidermal appendages is inexact and insufficiently descriptive. Simple means of preparation (whole mounts) can often provide taxonomists with much information that can be integrated with the precise usage and wealth of descriptive data the plant anatomist can offer.

Trichome types

Several trichome types have been the topics of monographs. Within the tribe Cichorieae of Compositae, genera are characterized by trichomes that serve as termini for laticifers (Zander, 1896). Bachman (1886) offers systematic and other data on peltate hairs. Glandular trichomes have been the subject of various studies, such as the early one of Martinet (1872). Description of glandular trichomes of insectivorous plants and citation of earlier literature on

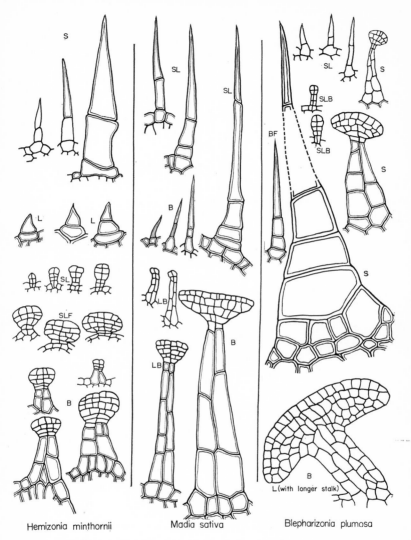

Hemizonia minthornii   Madia sativa   Blepharizonia plumosa

Fig. 3-4. "Trichome complements" of five tarweeds (subtribe Madinae of Heliantheae, Compositae), showing types of trichomes present; progressively greater differentiation in terms of complexity and of specificity of organographic occurrence on a plant; and variation within a particular type of trichome. Nonglandular trichomes are shown above, glandular trichomes mostly below

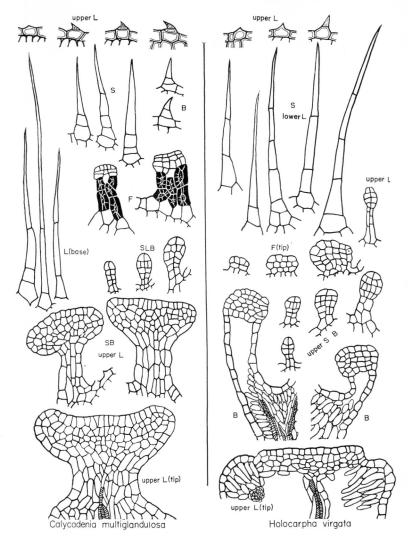

for each species. Organographic occurrence is indicated by abbreviations: S = stem; L = leaf; B = bract; F = flower (corolla). Special trichome types include tack-shaped gland (*Calycadenia multiglandulosa*, below), sessile gland (*Holocarpha virgata*, bottom), and hollow-stalked trichome (*Holocarpha virgata*, near bottom). All figures ×125.

this topic is given by Fenner (1904). Sessile glandular trichomes occur in a number of groups, such as Sparganiaceae and Typhaceae (Solereder and Meyer, 1933), and in this particular instance provide a link between these families.

Taxonomic use of trichomes

Extremely few angiosperms are truly glabrous. Most "glabrous" angiosperms are so because of degeneration of trichomes. This may in itself be of importance. For example, in *Hesperomannia* (Carlquist, 1957), trichomes in various species are identical on primordia, but relative preservation of uniseriate trichomes provides a specific characteristic.

A number of excellent studies attest the systematic importance of trichomes. Perhaps the most startling of these is that of Cowan (1950), which should be read by any student seriously interested in systematics. Cowan shows that in *Rhododendron*, trichomes are excellent criteria on subgeneric and specific levels, and contribute greatly to ideas on phylesis within the genus. Heintzelman and Howard (1948) show that types of trichomes, relative number of types present in a genus, and organographic distribution provide good criteria within Icacinaceae. For example, they find that segregation of *Ottoschulzia* from *Poraqueiba* is fully supported by evidence from trichomes. Similarly, Goodspeed (1954) finds that types and distribution of trichomes are correlated with specific and subgeneric distinctions in *Nicotiana*. A glance at the drawings of Metcalfe and Chalk (1950) for trichomes in *Croton* (Euphorbiaceae) will suggest their potential value in that genus. Trichomes may sometimes be used to evaluate ideas on relationship between families, as the work of Chorinsky (1931) indicates.

Methods of comparative study

Unfortunately, some comparative studies on trichomes suffer from incompleteness in several respects. Using examples from his studies on Madinae (1958b, 1959a, 1959b), the writer would like to call attention to certain types of representation that are attempted in Fig. 3-4. These comparisons utilize one species from each of five genera. The ideas employed in these comparisons are as follows: (1) Selection of all the trichome types—the *trichome complement*—rather than isolated trichomes is recommended; many angiosperms have more than one trichome type, and in the case of Madinae, a uniseriate nonglandular and a biseriate glandular type have been basic to the development of all trichomes. (2) Each trichome type in each species shows some variability, and thus each type should be studied as a *population* having extremes and a typical condition. (3) The *systematic distribution* of each trichome type should be given; in the case of Madinae, summary of such occurrence is given

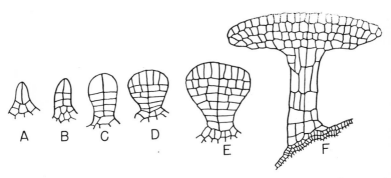

Fig. 3-5. "Ontogeny recapitulates phylogeny" in glandular trichomes of *Blepharizonia plumosa*. Note that earlier stages are identical in cellular constitution with the simplest type of glandular hairs found in Compositae (see Fig. 3-4, *Hemizonia minthornii*, middle); D is a stage that corresponds to the mature state of the most complicated trichome of *Hemizonia minthornii*; E is a stage comparable to the most complicated trichome of *Madia sativa*; F, the mature stage, is more complex than the trichomes in either of the two species mentioned. (Drawn from Carlquist, 1958.)

in Carlquist (1958b). (4) *Organographic distribution* of trichome types should be given. (5) Where trichomes within a phylad show stages in increasing complexity or reduction, *ontogenetic studies* (Fig. 3-5) should be utilized. In Madinae, such studies were necessary to establish that several peculiar glandular structures were, in fact, modified trichomes. By means of the five principles above, study of trichomes in a particular group can be put on a more precise basis.

Phylogeny of trichomes

Within a given group, trichomes show phylogenetic trends. The examples in Madinae (Fig. 3-4) show that from a simple biseriate glandular trichome, capitate types (multiseriate head portion) have been derived; these, in turn, have given rise to hollow-stalked glands, tack-shaped glands, and sessile glands. Examples of studies involving a phylogenetic viewpoint include papers by Cannon (1909), Cowan (1950), and Federowicz (1915). Netolitzky (1932) has offered a series of dicta on the phylogeny of trichomes that may be summarized as follows: (1) Trichomes have originated from papillate epidermal cells. (2) A one-celled trichome, unless shown to be a reduction, is more primitive than a several-celled hair; in the latter types, differentiation among the various component cells (for example, a glandular trichome) for special functions can occur. (3) Trichomes without radial symmetry or with mechanisms for becoming upright with relation to the surface on which they are borne are derived from those with radial symmetry or those that originate perpendicular to the surface on which they are borne. (4) If cells of a trichome acquire contents different from those of surrounding epidermal cells, this should be regarded as a specialization. (5) Every epidermal cell has the potentiality of becoming a trichome; this potentiality is affected by various influences. (6) Cells adjacent to a trichome may participate in the specialization of a trichome if they acquire structure and cell contents resembling those of the trichome but differing from those of ordinary epidermal cells.

These six criteria seem logical, but will require interpretation in individual cases. A corollary of the principles mentioned earlier in this chapter is that evolution of trichomes in a phylad is the sum of evolution in all trichomes. Thus evolution of the entire trichome complement should be studied, not merely one particular type.

Trichomes in analysis of hybrids

Because trichomes are often species characteristics, their usefulness to analysis of hybrids is considerable. Instances worth study are represented by the papers of Cannon (1909), Goodspeed (1954), Heiser (1949), and Rollins (1944).

REFERENCES

BACHMANN, O., 1886. "Untersuchungen über die systematische Bedeutung der Schildhaare," *Flora,* **69:** 387–400 *et seq.*

BAILEY, I. W., and NAST, CHARLOTTE G., 1944. "The comparative anatomy of the Winteraceae. V. Foliar epidermis and sclerenchyma," *Jour. Arnold Arb.,* **25:** 342–348.

CANNON, W. A., 1909. "Studies in heredity as illustrated by the trichomes of species and hybrids of *Juglans, Oenothera, Papaver,* and *Solanum,*" *Carnegie Inst. Wash. Publ.,* **117:** 1–67.

CARLQUIST, S., 1957. "Systematic anatomy of *Hesperomannia,*" *Pac. Sci.,* **11:** 207–215.

————, 1958a. "Anatomy of Guayana Mutisieae. Part II," *Mem. N. Y. Bot. Gard.,* **10:** 157–184.

————, 1958b. "Structure and ontogeny of glandular trichomes of Madinae (Compositae)," *Amer. Jour. Bot.,* **45:** 675–682.

————, 1959a. "The leaf of *Calycadenia* and its glandular appendages," *Amer. Jour. Bot.,* **46:** 70–80.

————, 1959b. "Glandular structures of *Holocarpha* and their ontogeny," *Amer. Jour. Bot.,* **46:** 300–308.

————, 1960. "Anatomy of Guayana Xyridaceae: *Abolboda, Orectanthe,* and *Achlyphila,*" *Mem. N. Y. Bot. Gard.,* **10:** 65–117.

CHORINSKY, F., 1931. "Vergleichend-anatomische Untersuchung der Haargebilde bei Portulacaceen und Cactaceen," *Öst. Bot. Zeit.,* **80:** 308–327.

COOPER, W., 1922. "The broad-sclerophyll vegetation of California," *Carnegie Inst. Wash. Publ.,* **319:** 1–124.

COWAN, J. M., 1950. *The* Rhododendron *leaf. A study of the epidermal appendages,* London: Oxford and Boyd.

DAMM, O., 1901. "Ueber den Bau, die Entwicklungsgeschichte und die mechanischen Eigenschaften mehrjähriger Epidermen bei den Dicotyledonen," *Beih. Bot. Centr.,* **11:** 219–259.

FEDEROWICZ, S., 1915. "Die Drüsenformen der Rhinanthoideae-Rhinantheae," *Bull. Int. Acad. Sci. Cracovie,* Math.-Nat. ser. B: 286–322.

FENNER, C. A., 1904. "Beiträge zur Kenntnis der Anatomie, Entwicklungsgeschichte, und Biologie der Laubblätter und Drusen einiger insectivoren," *Flora,* **93:** 335–434.

FRANCEY, P., 1936. "Étude de l'appareil stomatique chez les dicotylédones dans un but taxonomique," *Bull. Soc. Vaud. Sci. Nat.,* **59:** 1–12.

GOODSPEED, T. H., 1954. *The genus* Nicotiana, Waltham, Mass.: Chronica Botanica.

HEINTZELMAN, C. E., Jr., and HOWARD, R. A., 1948. "The comparative morphology of the Icacinaceae. V. The pubescence and crystals," *Amer. Jour. Bot.,* **35:** 42–52.

HEISER, C. B., 1949. "Study in the evolution of the sunflower species *Helianthus annuus* and *H. Bolanderi,*" *Univ. Calif. Publ. Bot.,* **23:** 157–208.

KEAN, C. I., 1931. "Light receptors in *Mesembryanthemum,*" *Trans. Bot. Soc. Edinburgh,* **30:** 37–42.

KOEHNE, E., 1884. "Über Zellhautfalten in der Epidermis von Blumenblättern und deren mechanischen Funktion," *Ber. Deutsch. Bot. Ges.,* **2:** 24.

LINSBAUER, K., 1930. "Die Epidermis," in K. Linsbauer, ed., *Handbuch der Pflanzenanatomie,* **4:** vii + 277.

36 · Epidermis and Trichomes

MARTINET, M. J., 1872. "Organes de secretion des végétaux," *Ann. Sci. Nat. Bot.,* sér. 5, **14:** 91–232.

METCALFE, C. R., and CHALK, L., 1950. *Anatomy of the dicotyledons,* Oxford: Clarendon Press.

MILDBRAED, J., 1908. "Stylidiaceae," in Engler & Prantl, *Das Pflanzenreich,* Leipzig: Verlag Engelmann.

MUNZ, P. A., and LAUDERMILK, J. D., 1949. "A neglected character in western ashes," *Aliso,* **2:** 49–62.

NETOLITZKY, F., 1932. "Die Pflanzenhaare," in K. Linsbauer, ed., *Handbuch der Pflanzenanatomie,* 4(4): 1–253.

PORSCH, O., 1905. *Die Spaltöffnungsapparat im Lichte der Phylogenie,* Jena: G. Fischer.

PRAT, H., 1932. "L'épiderme des graminées. Étude anatomique et systématique," *Ann. Sci. Nat. Bot.,* sér. 10, **14:** 119–324.

RACIBORSKY, M., 1895. "Die Schutzvorrichtungen der Blütenknospen," *Flora,* **81:** 151–194.

REHFOUS, L., 1923. "Sur la phylogénie des stomates," *C. R. Soc. Hist. Nat. Genève,* **40:** 68–78.

ROLLINS, R. C., 1944. "Evidence for natural hybridity between guayule (*Parthenium argentatum*) and mariola (*P. incanum*)," *Amer. Jour. Bot.,* **31:** 93–99.

SMITH, G. E., 1935. "On the orientation of stomata," *Ann. Bot.,* **49:** 451–477.

SOLEREDER, H. 1908. *Systematic anatomy of the dicotyledons* (trans. by Boodle and Fritsch), Oxford: Clarendon Press.

————, and MEYER, F., 1933. Systematische Anatomie der Monokotyledonen. Pandanales—Helobieae—Triuridales. 1(1): 1–155. Berlin: Gebrüder Borntraeger.

STAUDERMANN, W. von, 1924. "Die Haare der Monokotylen," *Bot. Arch.,* **8:** 105–184.

WARNEKE, F., 1911. "Neue Beiträge zur Kenntnis der Spaltöffnungen," *Jahrb. Wiss. Bot.,* **50:** 21–66.

ZANDER, R., 1896. "Die Milchsafthaare der Cichoriaceen," *Bibl. Bot.,* **37:** 1–44

chapter four ► Xylem

Because xylem of flowering plants exhibits such clear evolutionary trends in a variety of features, a consideration of the variation patterns in xylem cannot be divorced from discussion of phylogenetic trends. Indeed, the number of xylem characteristics is considerable (Tippo, 1946). Some features of xylem show what may be termed "major trends" of evolution in angiosperms and are based upon the vessel element and its evolution. Other features, such as septate fibers, represent specializations that appear to have occurred sporadically in various families, without reference to the major trends. Finally, there are a number of features, such as tyloses or crystals, that seem to bear little or no relation to the degree of specialization of wood in which they occur.

THE VESSEL ELEMENT: KEY TO XYLEM EVOLUTION IN ANGIOSPERMS

De Bary was probably the first to state clearly that a vessel is a vertical series of tracheids in which the pit membranes have been lost in pits of overlapping walls ("end walls"). The realization that vessel elements are phylogenetic derivatives of tracheids opened the way for development of ideas concerning vessel-element evolution. Interestingly, the most successful probes of this topic have involved use of statistical surveys of xylem characteristics, both in monocots and dicots. Very significant in this regard was the reconnaissance of Bailey and Tupper (1918) into length of tracheary elements in vascular plants and that of Bailey (1920a) into length of fusiform

37

initials. These studies demonstrated that (1) in general, vascular cryptogams have very long tracheids; (2) gymnosperms possess tracheids of intermediate length; (3) vessel elements in angiosperms are shortest; and (4) in those groups with secondary growth, length of tracheids or vessel elements is predetermined by length of fusiform cambial initials, and thus is an inherent, deep-seated feature of woody species.

The primitive vessel element

Investigations by others on the nature of the primitive vessel element in angiosperms and on trends of evolutionary specialization followed Bailey's work during the next three decades. Other studies showed how xylem features other than those of vessels are related to vessel evolution. These latter features are discussed below under appropriate categories. Before discussing them, however, we should examine the "morphological equations" upon which all of these studies rest—namely, those that apply to the vessel element itself.

Bailey and Thompson (1918) and Frost (1930a), working with dicots, and Cheadle (1943a), working with monocots, claim that the vessel element can have arisen phylogenetically from no other cell type than a tracheid. The evidence for this hypothesis depends upon the great morphological similarity of primitive vessel elements to tracheids in the woods where they occur. For example, in the fern *Pteridium,* the transition between tracheids and vessel elements in xylem is a subtle and almost imperceptible one. We would expect that if tracheids and their derivatives have shortened phylogenetically during advancement, primitive vessel elements would be longest, and if any angiosperms were primitively vesselless, their tracheids would be longer than most vessel elements. As Bailey and Tupper (1918) show, there is such a correlation, reinforcing the assumption (Fig. 4-1). Such primitive vessellessness has been established for Winteraceae, Tetracentraceae, Trochodendraceae, Amborellaceae, and one genus of Chloranthaceae, *Sarcandra.* It has been claimed by Takhtajan (1959) for Nymphaeales. No monocots have been shown to lack vessels primitively, although some (Araceae) have vessel elements difficult to distinguish from tracheids. If the assumptions of vessel-element origin are correct, we can expect that the following tracheidlike features in vessel elements would be primitive:

(1) great length, with comparatively narrow width; (2) long, overlapping end walls, which form a small angle with the vertical; (3) scalariform bordered pits (like those of fern tracheids) on lateral walls; (4) scalariform perforation plates with many bordered perforations (like pits on lateral walls except for loss of membranes); and (5) vessels angular in transectional view. These characteristics should show

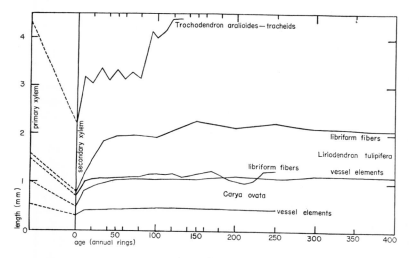

Fig. 4-1. Variation of tracheary elements with respect to age of stem for a vesselless angiosperm (*Trochodendron*); an angiosperm with vessels, but relatively primitive wood (*Liriodendron*); and an angiosperm with relatively advanced wood (*Carya*). Further explanation in text. (Redrawn from Bailey and Tupper, 1918.)

a strong degree of statistical correlation with each other. Moreover, tracheids similar to such vessel elements should occur associated with vessel elements in species with primitive xylem. Progressively less tracheidlike elements should show less statistical correlation with the five features listed above. Cheadle (1943a) pointed up the evidence in this regard for monocots much as Frost (1930a) did for dicots. Table 4-1 lists data compiled by Cheadle (1943a) from his survey of monocots, in which average vessel-element length is shown in relation to inclination of end wall and number of bars in the perforation plate.

Table 4-1

Length of Vessel Elements, mm

End-wall inclination	Perforation plates			
	Simple	1–10 bars	10–50 bars	Over 50 bars
Transverse	0.73	1.22	1.97	—
Slightly oblique	0.96	1.31	2.14	1.50
Oblique	1.15	1.69	2.41	3.44
Very oblique	—	—	2.56	4.22

This table clearly shows the strong correlation between long vessel elements, very oblique end walls, and perforation plates with many bars. The reverse conditions are also statistically correlated.

Morphological specialization in vessel elements

With establishment of the primitive characteristics of vessel elements in angiosperms, trends of advancement were studied. This was done for dicots by Frost (1930b, 1931) and for monocots by Cheadle (1943a, 1943b). These trends may be considered under five categories, corresponding to the five features listed above as exemplifying the primitive vessel element.

Length. As the data of Bailey and Tupper (1918) and Cheadle (1943a) show, there has been a tendency for phylogenetic decrease in vessel-element length in angiosperms corresponding to advancement. This tendency occurs not merely in secondary xylem but also in the primary xylem of dicots (Fig. 4-1) and monocots (Cheadle, 1943a).

End wall. The data of Frost (1930b) and Cheadle (1943b) give evidence for alteration of the end wall from a highly oblique to a nearly transverse angle (Fig. 4-2A-D).

Perforation plate. Variations in morphology of the perforation plate involve two features: (1) the loss of borders on the perforation plate (Fig. 4-3); and (2) the decrease of bars on the perforation plate (Fig. 4-2A-C) or other alterations of the bars (Fig. 4-2G-H) or even loss of the perforation plate and end wall (Fig. 4-2E-F). With regard to the first of these, Frost (1930b) has shown that loss of borders may take place in several ways, including loss from a simple perforation plate. In connection with loss of bars, as number of bars decreases, spaces between bars increase in size.

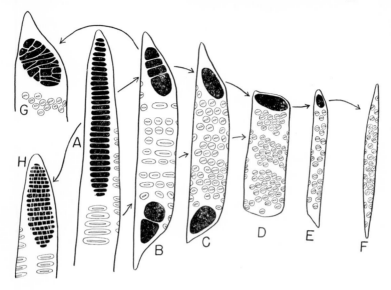

Fig. 4-2. Evolution of vessel elements of angiosperms. In A, a primitive, tracheidlike vessel with scalariform perforation plate is represented. Stages B and C show loss of bars on perforation plate. B-C-D form a series in tendency toward less oblique end wall. Steps E and F show how, by progressive narrowing, the perforation plate may be lost, resulting in a vascular tracheid. Examples G and H indicate alterations other than simplification that may occur on perforation plates. The lower series of arrows from A to D indicates evolution of lateral wall pitting, from scalariform to transitional, opposite, and finally alternate.

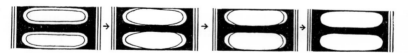

Fig. 4-3. Stages in loss of borders from perforations of a scalariform perforation plate. Although these stages were all observed in a single plant (*Hamamelis virginiana*), this trend has occurred within phyletic lines. (Redrawn from Frost, 1930b.)

The final stage in the evolution of the end wall—not included in the surveys of Frost and Cheadle—appears to be loss of the perforation, resulting in a "degenerate" vessel element, termed a "vascular tracheid." This phenomenon occurs only in specialized representatives of advanced families, such as Cactaceae (Metcalfe and Chalk, 1950) or Compositae (Carlquist, 1958). Vascular tracheids can be distinguished from true tracheids by the fact that where they occur, transitions to them from vessels with very small perforations (Fig. 4-2E) can be seen. Also, the highly advanced woods in which they occur otherwise lack tracheids or even fiber tracheids. The occurrence of anomalous perforations, such as those of *Echinocactus* and

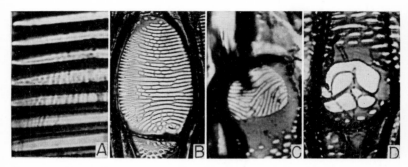

Fig. 4-4. "Anomalous" patterns of evolution in perforation plates of dicots. A, portion of a plate in *Hydrangea altissima*, showing vestigial borders and strands of wall material interconnecting the bars. B, *Carpenteria californica*, showing some forked bars and numerous prominent interconnections between bars. C, *Phoenicoseris berteriana*, plate showing alterations in orientation of bars and forking of bars. D, *P. regia*, a "multiperforate" plate, showing few bars dividing the plate into areas of various shape. B and D are diagrammed in Fig. 4-2, H and G respectively. A, ×500; B-D, ×320.

Mammillaria (Metcalfe and Chalk, 1950), are to be regarded as highly specialized conditions. Other types include reticulate perforations in which the bars form weak or strong interconnections (Fig. 4-4A, B), and reticulate plates in which bars form a series of forks or an irregular network (Fig. 4-4C, D). These bizarre perforations may represent modifications of scalariform perforation plates or retrogressions from simple plates (Thompson, 1923).

Lateral wall pitting. Lateral walls of vessels exhibit a similar series in specialization, showing the following evolutionary sequence in pitting type: scalariform, transitional, opposite, alternate (Fig. 4-2A-D, below). This trend, as the data of Frost (1931) show, applies both to intervascular pitting and vessel-ray pitting. Moreover, he finds a trend from fully bordered vessel-ray pit pairs to those that are half bordered (border on vessel side only) and finally to non-bordered pit pairs. If one compares the length of elements having scalariform perforation plates with the length of elements having scalariform lateral wall pitting, one finds that vessel elements with scalariform lateral wall pitting have a greater average length. Thus, if length of elements is a criterion, lateral wall pitting seems to have evolved more rapidly than perforation plates, because only the longer elements retain the scalariform lateral wall pattern (Frost, 1930b, 1931). Scalariform lateral wall pitting of another nature can be present in very advanced genera, particularly some genera with a "rosette-tree" habit, such as the Juan Fernandez Cichorieae (Carlquist, 1960a). This feature has been interpreted as a protraction of metaxylem lateral wall pitting into the secondary xylem.

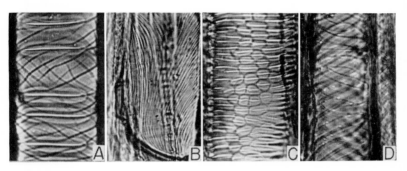

Fig. 4-5. Types of sculpture on walls of secondary xylem vessels. A, portion of a vessel from a primitive wood, *Hydrangea scandens*, showing helical bands on a wall bearing scalariform pits (appearance of crossed spirals is caused by spirals on wall of adjacent vessel); B, vessel of a specialized wood, *Wilkesia gymnoxiphium*, showing striae on a wall where pits are sparse; C, vessel from a specialized wood, *Hymenoclea salsola*, showing interconnection of adjacent pit apertures by grooves; D, vessel of an advanced wood, *Flotovia leiocephala*, showing prominent helical bands similar to those in A. All × *ca.* 300.

Another feature of walls of vessel elements in secondary xylem that has been interpreted phylogenetically is the occurrence of helical bands ("tertiary" helical thickening) or other helical sculpturing (Fig. 4-5). This wall relief may take the form of grooves linking pits in a helix (Fig. 4-5B) as well as bands (Fig. 4-5A, D) or striae (Fig. 4-5C). Spiral bands have been compared statistically to perforation-plate types by Metcalfe and Chalk (1950) who find that this feature occurs with somewhat greater frequency in woods with more primitive perforation plates, although the difference in frequency is not very great. Despite this lack of correlation with major trends of xylem evolution, a number of workers (Frost, 1931; Tippo, 1938; Carlquist, 1957, 1958) have considered spiral sculpturing an advanced phenomenon. Their contention is based upon the advanced nature of woods in which bands occur. An explanation for these divergent opinions on the phylogenetic interpretation of this feature may lie in the fact that "helical sculpturing" represents several phenomena that originate independently in various groups. Thus spirals (or grooves or striae) may occur in primitive as well as advanced groups. Within a particular group, presence of spirals may well be a minor specialization.

Transectional outline. The final feature of vessel elements that shows evolutionary progression is the shape of vessels as seen in transverse section. Cheadle (1943a) shows that in monocots there has been a transition from vessels angular in outline to vessels with even walls that are more nearly circular in outline. Frost (1930a) also reached this conclusion from his survey of dicots. Important exceptions to this trend occur in some highly advanced groups (Bailey, 1957a; Carlquist, 1958) where the angular characteristic may be attained secondarily.

Organographic and ontogenetic specialization of the vessel

Frost (1930a) has noted that transitions from tracheids to tracheidlike vessel elements (involving, for example, loss of borders on bars of a perforation plate) rarely occur in secondary xylem. Such transitions, however, can more often be observed in primary xylem, particularly metaxylem. These observations find an explanation in the fact, well known to wood anatomists, that primary xylem tends to have more primitive expressions of xylem characteristics than does secondary xylem. Thus, primary xylem is a sort of "refugium" of primitive expressions (Bailey, 1944a). As Bailey points out, the simple perforation plate is tardily attained by spiral-banded primary xylem vessel elements. The primitive type of perforation plate in such elements, a scalariformlike type, occurs as a series of bands narrowed across the end wall (Fig. 4-6A). Successively more advanced stages, showing loss of bands across the plate, are shown in Fig. 4-6B-E. Simultaneously with this introduction, as it were, of advanced vessel characteristics from secondary xylem into primary xylem in dicots, there is a tendency toward reduction in abundance of the primitive tracheary element, the tracheid, in the primary xylem. This trend parallels—although it lags behind—the similar trend in secondary xylem. These facts, together with the longer length of primary xylem elements as compared with those of sec-

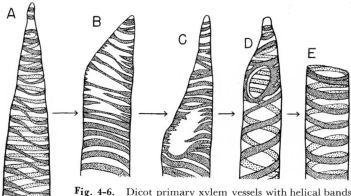

Fig. 4-6. Dicot primary xylem vessels with helical bands, showing progressive simplification of the perforation plate, like that shown for secondary xylem vessels in Fig. 4-2, A-D. Because vessels originated in secondary xylem of dicots, formation of advanced vessel types in primary xylem lags behind that in secondary xylem. (Redrawn from Bailey, 1944a.)

ondary xylem (Fig. 4-1), all contribute evidence for the hypothesis that vessels originated in the secondary xylem, then spread to primary xylem.

With reference to organographic origin of vessels in dicots, Bailey (1944a) shows that vessels originated in the secondary xylem of stems and roots. He states that "the origin and specialization of vessels in stems and roots appear in general to be closely synchronized, at least in the secondary xylem," whereas "there may prove to be evolutionary lags in leaves, floral appendages, and seedlings."

Although phenomena of ontogenetic and organographic origin of vessels were first discovered in dicots, the monocots, which lack secondary xylem, offer similar yet contrasting patterns. The data of Cheadle (1944) clearly show that within an organ, the late metaxylem is most advanced, and that more primitive tracheary-element characteristics occur in progressively earlier formed portions of the xylem. Although this accords with data from dicots, organographic origin of vessels in monocots offers a quite different picture. Cheadle's studies (1942, 1943a, 1943b, 1944) show that origin of vessels in monocots must have taken place in roots and progressed upward in the plant body because (1) most monocots possess vessels in roots only, with progressively fewer species showing vessels present in stems, rhizomes, inflorescence axes, and leaves, in that order, and (2) the direction for specialization in vessel elements follows the same organographic path. So decisive are these trends that Cheadle (1956) believes that, "given the information on vessels in the stem of a plant, one can predict without exception the limitations of specialization in the remainder of the plant." Such levels of specialization (Cheadle, 1955) may be related to the taxonomic system in a group of monocots. Note should be made of the fact that sequences in vessel evolution do not proceed at a uniform pace. Cheadle (1944) has shown that organs of monocots that have been tardy in acquiring vessels may, in some cases, show accelerated specialization of the vessel elements themselves.

Differential distribution of vessels in monocots and dicots—both organographically and in primary versus secondary xylem—forms a strong argument for independent origin of vessels in monocots and dicots, a subject explored by Cheadle (1953).

Perhaps the most significant single conclusion reached by Bailey concerning the origin and evolution of the vessel in angiosperms is that these developments have been irreversible. As such, vessel evolution, together with the correlative evolution of other xylem features, constitutes the "major trends of xylem evolution." These trends are unique in their apparent irreversibility. The latter is substantiated by the following evidence: (1) The vessel element can only have been derived from the tracheid. (2) Therefore, vessellessness is primitive (excepting cases of vessel loss in highly specialized plants). (3) The less tracheidlike a vessel element, the more advanced it is. (4) Control of vessel-element length in dicots is deep-seated, because it is governed by the length of fusiform cambial initials. (5) Progressive introduction of vessels into organs of a plant and into primary xylem from secondary xylem has been orderly and not random, and has been followed by similar progressive patterns in introduction of more and more advanced vessel types into these parts. (6) The evolution of the vessel is not an isolated phenomenon in the xylem, but has been correlated with a greater division of labor; that is, while vessels become less tracheidlike in secondary xylem, tracheids evolve into fiber tracheids and are finally replaced phylogenetically by libriform fibers (see below). (7) Various increasing specializations of a vessel element are statistically correlated, whereas if vessel evolution were reversible, one would not expect this. (8) Related to vessel-element evolution are a whole series of evolutionary trends (for example, in rays) in tissues that bear no direct relation to vessels. If reversion of vessel elements occurred, we would expect an orderly reversion of these trends, also. Bundle types in monocots (see Chapter 8) are an example of characteristics statistically correlated with vessel evolution. If reversion of vessel elements took place, features of tissues not directly associated with vessels would have to revert in a similar manner. (9) There have been no instances demonstrated in which vessel evolution has "gone backwards." For example, the apparent "tracheids" in a highly advanced wood that should, on the basis of its specialization, lack tracheids, can be demonstrated to be vascular tracheids. (10) Trends of tracheary-element evolution may be traced back into the geologic record. Bailey and Tupper (1918) and Chalk (1937) find that generally more primitive elements occur in fossil plants.

Plant anatomists are justly proud of the fact that the "major trends of xylem evolution" have been established independently of phyletic trends hypothesized by systematists working chiefly on the basis of floral morphology. There are signal points of agreement between the two, although the levels of specialization in xylem may show some deviation from those of floral morphology. In plants where such deviations occur, neither type of evidence should be neglected, and the possibilities of divergent rates of evolution in various parts of a plant should be explored. As stated by Bailey (1944a, 1957a), evidence from xylem is perhaps most useful in negating the derivation of a group with primitive xylem from one with advanced xylem characteristics. It seems self-evident that the well-established trends of xylem evolution should be taken into account by any systematist interested in formulating relationships or evolutionary status of angiosperm taxa. Caution must be used in applying the major trends of xylem evolution within a group, however, because these trends have been established from dicots or monocots) *at large*. As emphasized by Bailey (1957a), generalizations based on statistical methods tend to obscure, or override, lesser fluctuations or parallel evolution within a particular group. Influence of the habitat, source of the sample, age of the plant, and habit of the species may alter both quantitative and qualitative features. For example, if two species differ in vessel-element length, the one with longer elements is not necessarily more primitive.

SECONDARY XYLEM

The evolution of the vessel element has been discussed separately above because many patterns in this evolution are common to both dicots and monocots, and because it forms the keystone for interpreting evolution of other xylem features. In secondary xylem, there are an almost infinite variety of features that offer opportunities for evolutionary and taxonomic interpretation. In this section, then, taxonomic and evolutionary potentialities and uses of features other than those mentioned above are described. In addition, methods of collecting data, assessing results, and expressing evidence are reviewed for the various features of dicot wood. The sequence below follows that suggested by Tippo (1941). The usage of terms follows the compendium of the International Association of Wood Anatomists (Committee on Nomenclature, 1957). This glossary is essential for all students of wood anatomy. Attention should also be called to a paper (Committee on Standardization . . . , 1937) that gives a basis for measurements of tracheary elements. Some authors have found the size categories of Chattaway (1932) useful. The value of measurements is discussed by Rendle and Clarke (1934a, 1934b) and is considered in detail in Chapter 1. Under each category below, the most comprehensive papers dealing with each feature are listed. The purpose of this survey is not to describe these features, but rather to suggest the type of variation patterns and systematic value that may be expected. The original papers cited, or texts on wood anatomy may be consulted for illustrations. Also worthy of consultation are the lists given by Boureau (1957), Metcalfe and Chalk (1950), and Solereder (1908) of families that possess particular features.

Growth rings

Presence of growth rings and their width may be used systematically. Unfortunately, little is known in most species about whether growth rings are obligate, and, if they are not, what factors can induce them, or how much variability in ring width may be expected. The compilations of Chowdhury (1939–1940) are noteworthy.

Imperforate tracheary elements

The diversification of imperforate tracheary elements is one of the major trends of wood evolution, closely allied with the evolution of the vessel element. As summarized by Bailey (1953), there has been a tendency for increasing division of labor in the axial portion of the xylem. In vesselless plants, tracheids serve both for conduction and for mechanical support. Concomitant with the development of the vessel element into a structure efficient at conduction but less suited (on account of greater width, shorter length, and moderate wall thickness) for support, tracheids in woods possessing vessels have evolved features assuring greater mechanical support but less conduction. These features of imperforate elements that show evolution in the direction of greater mechanical support include (1) reduction in size and number of pits. Concomitant with reduction in size of pit membranes, a reduction in size of pit borders occurs. (2) increase in length of fibers as compared with vessels. Because longer, narrower cells are excellent for mechanical support, increase in length of fibers as compared with vessel elements has occurred (Fig. 4-1). Although vessel elements do not elongate appreciably during derivation from a fusiform cambial initial, libriform fibers do (Chattaway, 1936). Length of libriform fibers thus becomes an important feature in evolutionary and taxonomic considerations. Comparing the peak of the curve for Fig. 4-8 and Fig. 4-10, right (p. 53), we note that libriform fibers tend to have a greater length than vessel elements in dicots at large. Finally, (3) there is a tendency for greater wall thickness in many species as imperforate cells evolve into libriform fibers. Stages in evolution of libriform fibers from tracheids are shown in Fig. 4-7A-D.

The graph of Metcalfe and Chalk (Fig. 4-8) offers suggested

length categories; for designation of wall width, the size classes of Chattaway (1932) may be used.

Size and shape of pits in tracheids and fibers have been used as taxonomic characteristics. Size and abundance of pits vary on

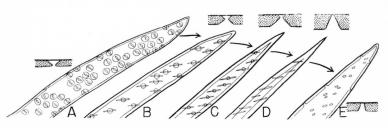

Fig. 4-7. Stages in phylogenetic change of tracheids in secondary xylem of dicots. Transectional outlines of pits are shown for each stage. A, tracheid with prominent bordered pits; B, C, fiber tracheids in which pit borders are diminished; D, a libriform fiber, showing disappearance of borders on pits; E, axial parenchyma cells, which (in certain cases *only*) are believed to have arisen from libriform fibers by means of fiber dimorphism. Evolution of fibers from tracheids often, as shown, involves narrowing of diameter of element and thickening of walls.

different walls, or in relation to the position of an element in a growth ring. Helical sculpturing, discussed above in connection with vessels, may be present on walls of tracheary elements other than

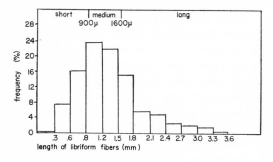

Fig. 4-8. Distribution of size classes of libriform fibers, based upon a large sampling of dicots. Suggested size designations are given above. (Redrawn from Metcalfe and Chalk, 1950, courtesy of Clarendon Press.)

vessel elements, as they are in some gymnosperms (for example, Taxaceae). The presence of "gelatinous" fibers or other phenomena related to "reaction" wood or "tension" wood deserve mention. For consideration of this field, see Boureau (1956), Wardrop and Dadswell (1955), or Rendle (1937).

The tendency of fibers to be septate or nonseptate has long been recognized as a useful feature for comparison of woods. However, the evolutionary interpretation of septation in fibers is not clear. Certainly statistical studies (Metcalfe and Chalk, 1950) show that fiber septation is not related to the major trends of xylem evolution.

Vessels

Abundance. Many wood anatomists have found that the number of vessels, as seen in a square millimeter of a transection, offers a characteristic of taxonomic significance. If vessels are infrequent, larger areas should be used. If vessels are difficult to identify in transection (because of similarity to vascular tracheids, for example), this measure becomes virtually impossible. For abundance classes, see Chattaway (1932). The possibility of phylogenetic increase (or decrease) in vessel abundance is a topic worthy of detailed investigation.

Groupings. Tippo (1946) shows solitary vessels as primitive, with groupings of vessels as an advanced expression. The figures of Metcalfe and Chalk (1950) on percentage of solitary vessels in woods of four classes of advancement support this view and place this feature among the "major trends of wood evolution." Thus, a figure for average number of vessels (pores) per group, as seen in transection, seems useful (Carlquist, 1957). Groupings can occur in various characteristic ways, such as pore multiples, pore clusters, and pore chains. Diagrammatic representations of these are given for many woods by Metcalfe and Chalk (1950).

Ring porousness. As opposed to random distribution of vessels in transection, many woods show large, numerous vessels at the beginning of a growth ring and fewer, smaller vessels near the end. Definitions of ring porousness are based upon differential distribution of vessels, but correspondingly, distribution of other elements may also be altered, and these data should be recorded as well. The development of ring porosity cannot be considered one of the "major trends of xylem evolution" because the data of Metcalfe and Chalk (1950) show that it is not definitely in correlation with features of vessel advancement. Gilbert (1940) claimed that ring porosity is an advanced feature, characteristic of some North Temperate woods. However, as Carlquist (1957) and others have shown, it certainly occurs in woods from the South Temperate Zone as well. Although it is clearly not appreciably modifiable in some species, the degree to which ring porosity is environmentally modifiable in any particular species needs investigation. Formation of ring porosity probably represents an evolutionary adjustment to highly seasonal climatic conditions that has occurred many times independently, in groups with primitive wood as well as in groups with specialized xylem. Certainly it is a taxonomically significant feature because of its restriction to a small percentage of temperate species.

Vessel-element size. The importance of vessel-element length in phylogenetic considerations has been stressed above, but one must remember that difficulties in obtaining representative data—for reasons described in Chapter 1—are very great. Chalk and Chattaway (1934) find that measurement of the total length of an element, rather than any part of it, gives the most reliable data. Macerated wood is the best source for such measurements, although many wood anatomists rely upon sections. As indicated by Stern and Greene (1958), quantitative features vary greatly with source of sample, and examination of Fig. 4-1 will show that stabilization in element length may be extremely slow. Provided that one has truly comparable material, simple statistical treatment (range in length, mean, and most frequent range) may be used. If truly reliable data are available, graphic representations (Fig. 4-9) such as those of Rodriguez (1957) are recommended. Distribution of vessel-element length and size classes are given by Metcalfe and Chalk (1950) in a table reproduced here as Fig. 4-10 (right).

Vessel diameter is quite significant in some angiosperms, although it may not be related to vessel-element length. There is a definite trend in some groups toward widening of vessels with increasing specialization. This may be accelerated in vines or lianas. Exceptionally narrow vessels are probably also specialized. The histograms of Metcalfe and Chalk give an idea of the range, abundance, and recommended size-class designations of these diameters (Fig. 4-10, left). Vessel diameter is most conveniently measured in transections, and should include the wall. The thickness of the wall itself can be an item of some importance, and Moseley (1948) found that this feature characterized species of *Casuarina*.

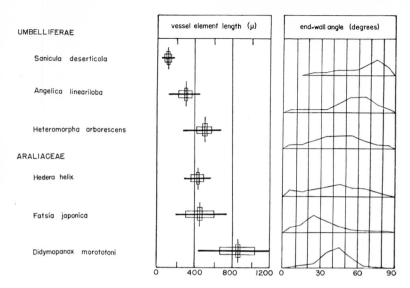

Fig. 4-9. Graphic methods of representing quantitative features of vessel elements. In "glyphs" representing element length, left, horizontal bar = range of length, vertical bar = mean, and boxes represent standard deviation and standard error; at right, end-wall angle is represented by a curve based on frequency of vessels with the various degrees of end-wall angle. If decreasing length indicates advancement, shift of glyphs further left represents advancement; if more nearly horizontal end walls are specialized, peak for curve farthest right represents advancement. (Redrawn from Rodriguez, 1957.)

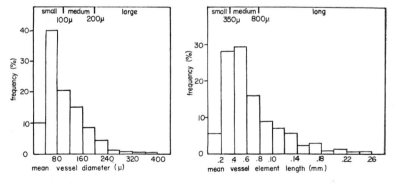

Fig. 4-10. Distribution of size classes for vessel diameter (left) and vessel-element length (right). Conventions as in Fig. 4-8. (Redrawn from Metcalfe and Chalk, 1950, courtesy of Clarendon Press.)

End wall. As discussed above, the end wall varies with respect to its angle and the nature of the perforation plate, and wood anatomists commonly discuss these features separately. The end wall may be measured as an angle with the horizontal or with the vertical, and expressed as an average or a range, or graphically (Fig. 4-9, right). Because the nature of the perforation plate is a sensitive index to the major trends of xylem evolution, much use has been made of it by various workers, and various classes have been invented. Such classes as "scalariform and simple plates" versus "simple plates only" or "scalariform plates only" have been used by Solereder (1908) and Moseley (1948). Rodriguez (1957) finds that such descriptions as "simple; 2 percent vestigial bars" suit his material, whereas Adams (1949) uses average number of bars, or "more than 30 bars" and "less than 30 bars" in Cornaceae. Presence of borders on perforations should be noted. Reticulate plates are not as rare as one might suppose (Chalk, 1933; Garratt, 1933a), and are worthy of description. Tippo (1941) suggests that width of perforations may be of value. Woodworth (1935) has called attention to what he terms "fibriform" vessel elements in Passifloraceae. These are distinctive not only in their elongate (longer than fusiform cambial initials) nature, but also in the anomalous placement of the small perforations, which occur well below the tip and suggest that apical elongation like that of fibers has occurred.

Finally, presence of vascular tracheids ("degenerate vessel elements") is an exceptionally interesting feature of certain highly advanced woods. The nature of vascular tracheids and concepts that permit one to distinguish them from true tracheids are described above in connection with vessel evolution. There is no question of their derivative nature; in Compositae, they appear to occur in the most advanced members of phyletic lines (Carlquist, 1958). Their formation may, to some extent, be related to aridity. A listing of families possessing vascular tracheids is offered by Boureau (1954).

Outline. The importance of angular or rounded appearance of vessels as seen in transection is discussed above in connection with vessel evolution.

Pitting. A distinction is customarily made between pitting between adjacent vessels (intervascular pitting) and that which occurs between a vessel and a parenchyma cell. This distinction is a useful one because, for example, intervascular pitting may be alternate in a vessel that has scalariform pitting on walls facing parenchyma. Solereder (1908) lists families in which large simple pits occur on vessel walls facing parenchyma as well as families in which only bordered pits occur on such walls. As mentioned above, loss of borders on vessel-parenchyma pitting has apparently taken place in dicots (Frost, 1931). More importantly, the evolution of lateral wall pitting (scalariform—opposite—alternate) is one of the more important of the major trends of xylem evolution.

According to Metcalfe and Chalk (1950), evolution of minute intervascular pitting represents a specialization in certain families, and is not related to the major trends of xylem evolution. Size and relative abundance of pits are used as taxonomic criteria by such workers as Record (1934), and Record and Chattaway (1939) have set up convenient classes for pit size. These classes apply only to opposite or alternate pitting. Intervascular pitting often differs in abundance from vessel-parenchyma pitting, and note is often made of this in anatomical descriptions.

A peculiar feature of a few dicot families is the presence of protuberances surrounding the pit apertures in pits of vessels (and fiber tracheids). If a species has such pits, known as vestured pits, they occur in vessels throughout the plant. Although especially thin sections are required to demonstrate vesturing, it is an excellent diagnostic criterion, and the reader is referred to Bailey (1933) for a full account. Presence of peculiar "crateriform" pits is a specific characteristic in *Cercidium* (Cozzo, 1953).

Sculpturing on vessel walls. Under this heading may be grouped the various forms of relief shown in Fig. 4-5. The phylogenetic significance of these has been discussed above in connection with vessel evolution. Sculpturing offers a useful diagnostic criterion, and the reader should consult the tabulations of Solereder (1908) and Metcalfe and Chalk (1950). Sculpturing of various types does not appear to be related to the major trends of vessel evolution, but studies are needed to establish the significance of these forms of relief.

Tyloses. The significance of tylosis formation is certainly a physiological one, related to inactivation of vessels and the continued activity of protoplasts of adjacent parenchyma cells that penetrate the vessels. This process, however, characterizes the wood of certain species only, and is apparently totally absent in some. Alten (1909) and Record (1934) have offered lists of genera in which they have been reported, and doubtless many more genera could be added. Tyloses are abundant in some taxa, occasional in others. In some species with tyloses, sclerification occurs, so that tyloses may become extremely thick walled. This feature, and presence of visible pits in tyloses, may form taxonomic criteria (for example, *Fitchia;* see Carlquist, 1958).

Vascular rays

Abundance. The number of rays present in an area of a section would be difficult to estimate. The simplest measure devised is the number of rays that intersect a 1-mm line running at right angles to the axis on a tangential section. A diagram giving distribution of ray abundance based on this measure is shown in Fig. 4-11

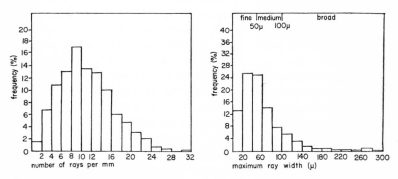

Fig. 4-11. Distribution of abundance classes (left) and width classes (right) for rays in dicot woods. Ray abundance is measured by the number of rays which intersect a linear mm across a tangential section. Ray width is also based upon tangential sections and is based upon averages of maximum width of each ray. (Redrawn from Metcalfe and Chalk, 1950, courtesy of Clarendon Press.)

(left). Abundance can also be measured in transections. Abundance classes are given for this measure by Chattaway (1932). Ray abundance is a feature often included in keys because of its taxonomic value; in particular groups, phylogenetic tendencies are worthy of exploration.

Dimensions. Dimensions of rays, as seen in tangential sections, are often used for comparative purposes. Measurements are made on rays surrounded by fibers rather than by axial parenchyma. The suggested size categories and distribution of widths is given in the diagram shown in Fig. 4-11. Width of multiseriate rays as seen in tangential section can be expressed either in units or in number of cells at maximum width. Relative ray size is sometimes characteristic of taxonomic groups (for example, the large rays of Proteaceae; see Tupper, 1927).

Ray height is a difficult feature to use in some species because of tremendous variation. If this measure is feasible, uniseriate rays are measured separately from multiseriate. These figures can be expressed in terms of maximum, average, range, and most frequent range.

Histology. Anatomy of rays is considered in terms of two histological respects: (1) width (one cell or more than one cell wide) and (2) cellular composition. With respect to the first, rays are termed multiseriate if any portion is more than one cell wide and uniseriate if entirely one cell in width. A wood may have uniseriate rays exclusively, multiseriate rays exclusively, or both types. Complete absence of either one of the two types is fairly rare, but if one type is quite infrequent in a wood, it is generally disregarded. In regard to the second feature, composition of rays, wood anatomists recognize erect (upright, or vertically elongate), procumbent (radially elongate), and square (isodiametric) cells. These types can be observed accurately only in radial sections. Square cells are considered morphologically equivalent to erect cells. If procumbent cells only, erect and square cells only, erect cells only, or square cells only occur in a ray, it is said to be homocellular. If procumbent and erect, or procumbent and square cells occur in a ray, it is termed heterocellular. An additional feature of significance is provided by variations in length of uniseriate wings on multiseriate rays. The Committee on Nomenclature (1957) recommends that rather than designation of particular types, the various expressions described above be united into a sentence or phrase. For example, "the multiseriate portions composed of procumbent cells, the uniseriate margins composed of square to erect cells," or "rays uniseriate, composed entirely of procumbent cells."

Various designations, indicated parenthetically in Fig. 4-12, were used by Kribs (1935) for his comparisons between various types of ray occurrence in dicots and trends of vessel evolution.

Although Kribs' types have been used for comparisons in many papers, they are not now in general use. Nevertheless, his statistical comparisons show that ray evolution belongs to the major trends of xylem evolution. Simplified, his conclusions may be summarized as follows:

Rays in the most primitive woods exhibit the following features:

(1) Both multiseriate and uniseriate rays are present.
(2) Both types of rays are high (of marked vertical length).
(3) Both types of rays are heterocellular.
(4) Multiseriate rays have long uniseriate wings.

With advancement, the following changes in rays occur:

(1) Multiseriate rays or uniseriate rays are lost.
(2) There is a tendency for loss of heterogeneity in ray cells, particularly loss of erect cells. Thus, homocellular rays are advanced.
(3) Multiseriate rays are reduced in size and number.
(4) Uniseriate wings on multiseriate rays are reduced, ultimately to a single cell.

Barghoorn (1940, 1941a, 1941b) extended Kribs' work and showed some additional trends in advancement. Barghoorn clearly demonstrated that loss of heterogeneity in rays can result in homogeneous rays composed of erect cells only. These changes are the result of increase in length of ray initials. The ultimate result of continued elongation of ray initials, as shown by Barghoorn (1941b), is conversion of ray initials into fusiform initials. Thus, raylessness is the end result of this process. The evidence presented by Kribs and Barghoorn is presented in a semidiagrammatic form in Fig. 4-12.

Just as vessels show different degrees of specialization in a plant within earlier as opposed to later formed xylem, rays show similar alterations in their specialization during ontogeny of a stem. Basing their comments upon Barghoorn's findings, Bailey and Howard (1941b) offer the following statement:

> Thus, many dicotyledons tend to retain relatively primitive types of ray structures in the first-formed secondary xylem after the rays of the outer wood have become more or less extensively modified. On the contrary, reduction or elimination of multiseriate rays frequently progresses in an inverse direction, viz., from the earlier toward the

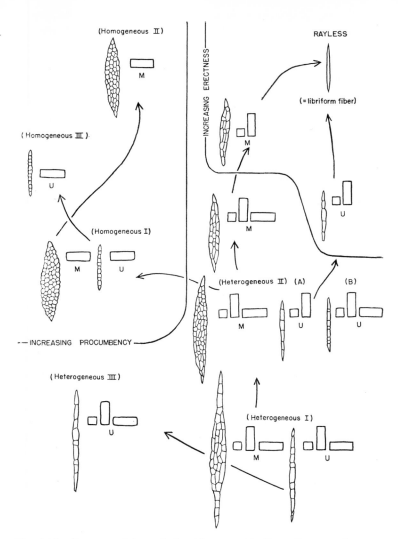

Fig. 4-12. Some trends in evolution of ray types in dicots. Rays are shown in tangential section; to the right of each, conventionalized shapes represent presence of procumbent or erect cells (as seen in radial section) for a multiseriate (M) or uniseriate (U) ray. Ray types are based upon those of Kribs, with modifications. Ray cells square as seen in radial section are considered morphologically equivalent to erect cells. (Based in part upon Kribs, 1935.)

later stages of the ontogeny of the secondary body. Under these circumstances, the outer secondary xylem exhibits more primitive types of ray structure than does the first-formed secondary xylem.

Apart from these major features of ray histology, other aspects deserve mention. The term "aggregate ray" has been applied to groupings of narrow rays, often uniseriates, separated from each other by imperforate elements only, and that in gross aspect appear to be a single ray. Aggregate rays are a diagnostic feature of certain woods (see listing in Metcalfe and Chalk, 1950). The presence of sheath cells (upright cells forming a sheath around the procumbent cells of a multiseriate ray) may also be a diagnostic characteristic. Tile cells (upright cells the same height as procumbent cells, in horizontal series) are characteristic of several families of Malvales (Chattaway, 1933). Perforated ray cells occur occasionally or characteristically in various dicots, a listing of which is given by Chalk and Chattaway (1933). These cells are vessel elements derived from ray initials and are somewhat intermediate between ray cells and vessel elements in morphology; for example, pits have reduced borders and are smaller than those on true vessel elements within a species (Carlquist, 1960b). Perforated ray cells appear frequently in woods in which ontogenetic breakup of rays is active.

Cell wall. Ray cells may be thin walled and nonlignified, particularly in some perennial herbs (Carlquist, 1958). Various degrees of lignification of ray cells occur, and such expressions are excellent systematic criteria. For a review of pitting among ray-parenchyma and axial parenchyma cells, with examples of taxonomic usefulness of these features, see Frost (1929).

Taxonomic usefulness of rays. The various features described above can be combined to offer excellent taxonomic criteria. For example, in *Argyroxiphium* (Carlquist, 1958), both multiseriate and uniseriate rays are present; rays are heterocellular and consist of thin-walled nonlignified cells. In a related genus, *Dubautia,* multiseriate rays are present exclusively, rays are homocellular, consisting of square to erect cells only, and ray-cell walls are thicker and lignified.

Axial parenchyma

Axial (vertical) parenchyma, which is derived from fusiform cambial initials, contains some of the most subtle as well as most conspicuous characteristics in wood anatomy. Often, axial parenchyma cells are not greatly different from fibers in shape and wall thickness. Parenchyma cells can be recognized by distribution on the transverse section, characteristics of pits in cell walls, and nature of cellular contents. Although wood anatomists conventionally work from dried material, liquid-preserved samples can be valuable in demonstrating living contents in parenchyma cells.

Types of cell distribution. Various types of parenchyma distribution are shown diagrammatically in Fig. 4-13. These types fall into two main groups: (1) apotracheal, in which parenchyma is distributed without specific relation to vessels and (2) paratracheal, in which parenchyma shows a close association with vessels. Good photographic representations of the various types are given by Metcalfe and Chalk (1950), Kribs (1937), and Hess (1950). Many diagrammatic views are given for various taxa in the main body of Metcalfe and Chalk (1950).

As in ray parenchyma, Kribs compared a series of levels of vessel specialization to these various axial parenchyma types (Kribs, 1937). His results are presented in Table 4-2 (terms changed to conform to recent usage):

Table 4-2

Percentages of Parenchyma Types in Each Vessel Type

Type of vessel element	Diffuse	Diffuse-in-aggregates	Para-tracheal scanty	Apotra-cheal banded narrow	Apotra-cheal banded wide	Vasi-centric	Absent	Ter-minal
Scalariform I	69.84	19.04	....	....	....	....	11.12	....
Scalariform II	59.37	15.62	....	....	....	....	12.50	12.51
Scalariform-porous	13.23	41.18	20.59	8.82	....	....	11.76	4.42
Porous-oblique	11.88	32.65	15.78	8.40	....	11.96	12.87	6.45
Porous-oblique to transverse	4.16	14.06	7.81	17.81	10.42	31.28	4.16	10.93
Porous, transverse	....	2.69	....	6.66	9.33	70.22	1.77	9.33

The results of these statistical compilations show that axial parenchyma has evolved in accordance with the major trends of xylem evolution and may be summarized as follows:

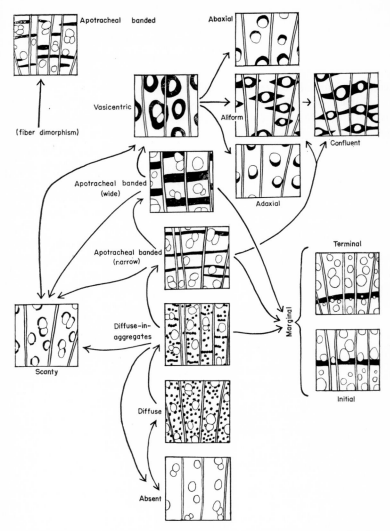

Fig. 4-13. Some trends in evolution of axial parenchyma distribution patterns in dicots. Squares represent diagrammatic transections; black areas = axial parenchyma. (Based in part upon Kribs, 1937.)

(1) Absence of parenchyma is primitive, at least in some cases; this seems confirmed, for example, by the absence of parenchyma in vesselless woods such as Winteraceae (Bailey, 1944b).

(2) Diffuse parenchyma represents the primitive type of parenchyma distribution. Expressed simply, fusiform initials are matured into parenchyma cells directly, in the same fashion as other derivatives become tracheids, etc.

(3) Diffuse-in-aggregates is slightly advanced over diffuse, and illustrates that a tendency toward grouping of cells often marks an advance.

(4) Apotracheal banded parenchyma represents the most advanced type of apotracheal parenchyma. Viewed statistically, wide bands appear slightly more advanced than narrow bands.

(5) Terminal parenchyma and absence of parenchyma are scattered among woods of varying degrees of specialization, and may be assumed to have arisen independently several times, concurrently with the major trends of axial parenchyma evolution. The relatively long vessel elements of woods in which parenchyma is absent suggest that elimination of, or primitive absence of, parenchyma is more characteristic of primitive woods than specialized ones. Terminal and initial parenchyma (which may be termed marginal parenchyma, collectively) may represent evolutionary modifications in response to particular ecological conditions. For a discussion of marginal parenchyma, see Chowdhury (1936).

(6) Of the paratracheal types, scanty appears to be less specialized than vasicentric, as a generalization.

One cannot convert Kribs' table into a simple phylogenetic scheme directly, and in a particular group, origins of axial parenchyma types may not be as indicated. For example, Bailey and Howard (1941a) note that additional trends must be recognized:

> (1) broad-banded apotracheal types arise from diffuse ones through various transitional narrow-banded apotracheal types. (2) in many families, vasicentric, aliform, and confluent types originate from banded types, and (3) through excessive reduction, scanty paratracheal types may arise at various levels of the differentiation of both apotracheal and paratracheal types. Furthermore . . . there are many complex types of parenchyma distribution which are transitional between typical apotracheal and purely paratracheal types.

Other exceptional cases are pointed out by Bailey (1957a), who states that parenchyma bands in Leguminosae may not be groupings of pre-existing parenchyma cells. Rather, they may have originated *de novo*. The writer (1958) illustrated how such origin may have taken place within a single genus, *Dubautia*, by a sort of fiber dimorphism (see Fig. 4-7D-E). Thin-walled fibers may become diversified to form two types: long, narrow, thick-walled fibers, and short, wide, thin-walled fibers, often septate with living contents. The latter, because of their histological characteristics, must be termed parenchyma cells. Thus, we can add a special case in which dimorphism produces parenchyma.

The above conclusions have been incorporated into Fig. 4-13. Following recent usage, both types of marginal parenchyma, initial as well as terminal (which probably formerly included initial), are recognized. These various types of parenchyma provide excellent diagnostic features. Descriptions of banded types should indicate an estimate of width of the band (preferably in number of cells), and similar figures are useful for paratracheal types. Many wood descriptions contain figures on the number of cells per strand—that is, the number of cells derived from each fusiform initial.

Cell characteristics. Kribs (1937) also compared axial parenchyma types with length and width of the cells themselves, and with vessel-element length in woods with the various parenchyma types. Interestingly, length and width of parenchyma cells are correlated with major trends of xylem evolution. Parenchyma cells tend to become shorter and wider in more specialized woods. This is due, at least in part, to horizontal subdivision of the fusiform initials into strands.

The nature of pitting in cell walls of axial parenchyma is often described, just as are similar features of rays. One may cite size, abundance, and distribution of pits, whether between adjacent axial parenchyma cells or between parenchyma and other cell types. Axial parenchyma cells, like ray cells, can vary in wall thickness and lignification, and these are often important diagnostic features.

Characteristics involving more than one cell type

Storied structure. The foundations for an understanding of storied structure of wood were laid by Bailey (1920b, 1923). He showed that storied structure depends on the plane of division of cambial initials. Initials in less specialized woods divide in a radial horizontal plane, with subsequent elongation of the initials, so that the two daughter cells overlap side by side, increasing the circumference of the cambium. In some specialized woods, on the other hand, cambial initials divide in a radial longitudinal plane, so that without apical elongation of the cells, cambial circumference is increased. Radial longitudinal divisions of cambial initials eventually produce tangential rows of initials, which thus give rise to storied elements in the secondary xylem. Storied structure occurs in woods that have short (and thus, advanced) tracheary elements and other specialized features. There seems little doubt that storying is specialized. Its taxonomic distribution (Solereder, 1908) along with other lines of evidence suggests that storying has originated many times independently in dicots.

Descriptions of storied woods should mention which elements are storied, and a partial survey in this regard is given by Boureau (1957). There is a range in storying from woods in which only the fibers may be vaguely storied to those in which rays, as well as all the axial elements, conform to the storied pattern.

Laticifers. Morphology and classification of laticifers in angiosperms at large is considered in Chapter 2. Instances of their occurrence in secondary xylem are considered taxonomic criteria by Ingle and Dadswell (1953) and Woodworth (1932).

Vasicentric tracheids. Short and irregularly shaped tracheids may be present adjacent to vessels, which often distort the tracheids by their lateral expansion. Presence of such tracheids is characteristic of certain families, a listing of which is given by Metcalfe and Chalk (1950). According to these authors, these tracheids are usually taxonomic criteria for tribes within those families.

Included phloem; anomalous stem structure. As the review of Chalk and Chattaway (1937) on these topics shows, included phloem (strips of phloem surrounded by secondary xylem) may be the product of a normal cambium. In other cases, it is produced by successive, isolated, or otherwise abnormal cambia. These modes of origin correspond to the types of included phloem "foraminate" and "concentric" given in the listings of Metcalfe and Chalk (1950). A relatively small proportion of dicots have woods with included phloem. It occurs in a scattering of genera from a variety of families. Included phloem undoubtedly represents a series of specialized conditions, whether in response to a particular growth habit or for some other reason. The diagrams, descriptions, and keys of Chalk and Chattaway (1937) are excellent for showing the diversity of types involved. For discussion of other papers dealing with anomalous stem structure, see Chapter 8. "Internal," or intraxylary, phloem is considered in Chapter 6.

Interxylary cork. A few dicots annually produce cork layers in the xylem, and in old stems these occur as concentric rings. This peculiarity is apparently most frequent in plants from xeric or halophytic environments and is probably a desiccation-resistance mechanism. For an example showing the systematic usefulness of this feature in *Artemisia,* see Moss (1940).

Secretions; intercellular spaces. Secretions of gums, resins, resinlike materials, and other substances occur in woods (1) in specially formed canals and cavities; (2) in inactive cells, such as heartwood vessels; (3) in intercellular spaces that are not differentiated as a canal or cavity; and (4) in various distributions as a result of traumata. Canals and cavities may occur in rays or axial tissue. For a review of terminology and literature on canals and cavities, see Stern (1954b). These structures may be taxonomically important. For example, vertical canals occur in Dipterocarpaceae and Vochysiaceae. Canals in rays characterize many Araliaceae, Umbelliferae, Anacardiaceae, and Burseraceae.

Disjunctive cells. In some woods, parenchyma cells or tracheids may become separated from each other, remaining in contact by means of elongated processes. These cells have been noted as diagnostic features by some wood anatomists.

Crystals. Crystal types and location of crystals in woods have been studied because in certain cases they can be useful systematically. A detailed review of crystal occurrence in wood is given in two papers by Chattaway (1955, 1956). Although crystals are best seen in sectioned material, they can be overlooked because of their translucent nature, and simple polarizing equipment is often helpful in locating them.

Starch grains. Starch grains have been reported in axial and ray parenchyma of certain woods. They have also been noted in septate "fiber tracheids" (Harrar, 1946).

MEANS OF COMPARISON

The above listing of characteristics suggests the great variety of comparative materials that xylem offers for taxonomic and evolutionary studies. The variety is, in fact, so great that one may well ask how such data should be collected or organized. Various means are available, but Rendle and Clarke (1934a) aptly protest against the mass of descriptive text that so often defies extraction of pertinent material. Cures for such obfuscation lie mostly in use of keys, tables, graphs, or other means of placing in relief those features that offer the best means of comparing taxa.

Keys. Although keys are, at least theoretically, primarily useful for identification of materials, they can also summarize considerable information in a neat and brief fashion. Many examples of keys could be cited. Among them the papers of Ingle and Dadswell (1953, 1956) and Record (1934) are noteworthy.

Tables. Tabular data can lend themselves to a variety of purposes, from simple summarization of qualitative and quantitative data to demonstrations of phylogenetic trends. In this latter regard, the table of Moseley and Beeks (1955, p. 325) is worthy of mention.

Descriptions. Although recent work has emphasized use of the above methods, such summaries do not necessarily replace textual descriptions. These may be given in terms of genera and species (Ingle and Dadswell, 1956; Shutts, 1960) or structural features (Carlquist, 1958). Most important, however, is the necessity for integrating wood anatomy with other types of data, a necessity that is being increasingly recognized in modern morphological studies.

ACHIEVEMENTS OF WOOD ANATOMY

Wood anatomists have, for the most part, been modest in advertising the taxonomic and even phylogenetic significance of their data. Such modesty, at least in part, derives from an awareness that these data do not provide the full body of evidence needed for solution of problems. Wood anatomy may, however, provide just that additional measure of evidence needed for solution of problems (Chalk, 1944; Vestal, 1940). In studies of genera of uncertain posi·

tion, wood anatomy proved invaluable to Stern and Brizicky (1958) for showing that *Heteropyxis* is myrtaceous. Likewise, Tippo (1940) was able on such bases to place the monogeneric family Eucommiaceae near Ulmaceae.

As one series of examples, one may cite the manner in which studies of wood anatomy have aided in collapsing the taxonomic concept "Amentiferae." These studies do not merely reject this concept, they also outline new natural relationships for the taxa concerned. Thus, Moseley (1948) delineates affinities between Casuarinaceae and Hamamelidaceae. Hall (1952) finds that Fagaceae may represent a rosalean derivative, and Stern (1952) demonstrates similarities between Julianiaceae and Burseraceae or Anacardiaceae. Tippo (1938) and Moseley (1948) find reason for allying Urticales with Hamamelidales, and Garryaceae appear to belong to Umbellales on the basis of such studies (Moseley and Beeks, 1955; Rodriguez, 1957). In addition to demonstrating the unnatural qualities of the "Amentiferae," these authors emphasize that these woods are not particularly primitive, considered separately or as a whole, so that we cannot consider any of the catkin-bearing families to approximate ancestral conditions in angiosperms on the basis of xylem. This is an example of the ways in which wood anatomy may offer "negations" (Bailey, 1957a). Certainly the evidence accumulated by the authors just cited concerning the relationships of the catkin-bearing families shows clearly that positive types of conclusions may be reached also.

During the renascence of interest in Ranales, anatomy of wood has been the topic of many studies. These studies clearly indicate that although many woody Ranales contain primitive features, many of the woods in this order are not as primitive as one would expect. Moreover, the amazing diversity of Ranales—to the point of being an unnatural order—is revealed. Studies of vesselless Ranales, such as *Trochodendron* and *Tetracentron* (Bailey and Nast, 1945), Winteraceae (Bailey, 1944b), *Sarcandra* (Swamy and Bailey, 1950), and *Amborella* (Bailey and Swamy, 1948; Bailey, 1957b), have shown that although diversity and degrees of modification occur in this group of genera, trends of wood evolution based upon vessel evolution are confirmed in these primitively vesselless groups. This contrasts with floral morphology, which is rather specialized in some of the genera (for example, *Sarcandra*). In a close alliance of three families, Degeneriaceae (Bailey and Smith, 1942), Magnoliaceae (Canright, 1955), and Himantandraceae (Bailey, Nast, and Smith, 1943), wood anatomy contrasts with the marked primitiveness of floral structure. Although *Degeneria* exhibits rather primitive xylem, *Galbulimina* (*Himantandra*) is rather advanced (for a ranalean family) whereas Magnoliaceae show a wide gamut of specialization. Scandent Ranales, such as *Austrobaileya* (Bailey and Swamy, 1949), and *Illicium, Schizandra,* and *Kadsura* (Bailey and Nast, 1948), have (especially in *Austrobaileya*) primitive types of wood modified with relation to the twining habit. Several families, although relatively advanced in floral features, possess primitive xylem: Eupteleaceae (Nast and Bailey, 1946), Cercidiphyllaceae (Swamy and Bailey, 1949), Eupomatiaceae (Vander Wyk and Canright, 1956), and Canellaceae (Wilson, 1960). Studies on wood anatomy have allied Hernandiaceae (Shutts, 1960), Gomortegaceae (Stern, 1955), Lauraceae (Stern, 1954a), and Monimiaceae (Garratt, 1934). These four families show many advanced xylem features. Likewise, the families Myristicaceae (Garratt, 1933a, 1933b) and Annonaceae (Vander Wyk and Canright, 1956) show relatively specialized features.

A number of additional instances in which wood anatomy has proved invaluable in solution of taxonomic problems are cited by Tippo (1946). Certainly the taxonomist of today can no longer overlook this type of evidence and its tremendous significance in achievement of the goals of systematic botany.

REFERENCES

ADAMS, J. E., 1949. "Studies in the comparative anatomy of the Cornaceae," *Jour. Elisha Mitchell Sci. Soc.,* **65:** 218–244.

ALTEN, H. VON, 1909. "Kritische Bemerkungen und neue Ansichten über die Thyllen," *Bot. Zeit.,* **67:** 1–23.

BAILEY, I. W., 1920a. "The cambium and its derivative tissues. II. Size variations of cambial initials in gymnosperms and angiosperms," *Amer. Jour. Bot.,* **7:** 355–367.

———, 1920b. "The cambium and its derivative tissues. III. A reconnaissance of cytological phenomena in the cambium," *Amer. Jour. Bot.,* **7:** 417–434.

———, 1923. "The cambium and its derivative tissues. IV. The increase in girth of the cambium," *Amer. Jour. Bot.,* **10:** 499–509.

———, 1933. "The cambium and its derivative tissues. VIII. Structure, distribution, and diagnostic significance of vestured pits in dicotyledons," *Jour. Arnold Arb.*, **14**: 259–273.

BAILEY. I. W., 1944a. "The development of vessels in angiosperms and its significance in morphological research," *Amer. Jour. Bot.*, **31**: 421–428.

———, 1944b. "The comparative morphology of the Winteraceae. III. Wood," *Jour. Arnold Arb.*, **25**: 97–103.

———, 1953. "Evolution of tracheary tissue of land plants," *Amer. Jour. Bot.*, **40**: 4–8.

———, 1957a. "The potentialities and limitations of wood anatomy in the phylogeny and classification of angiosperms," *Jour. Arnold Arb.*, **38**: 243–254.

———, 1957b. "Additional notes on the vesselless dicotyledon, *Amborella trichopoda* Baill.," *Jour. Arnold Arb.*, **38**: 374–378.

———, and HOWARD, R. A., 1941a. "The comparative morphology of the Icacinaceae. III. Imperforate tracheary elements and xylem parenchyma," *Jour. Arnold Arb.*, **22**: 432–442.

———, ———, 1941b. "The comparative morphology of the Icacinaceae. IV. Rays of the secondary xylem," *Jour. Arnold Arb.*, **22**: 556–568.

———, and NAST, CHARLOTTE G., 1945. "Morphology and relationships of *Trochodendron* and *Tetracentron*. I. Stem, root, and leaf," *Jour. Arnold Arb.*, **26**: 143–154.

———, ———, 1948. "Morphology and relationships of *Illicium*, *Schizandra*, and *Kadsura*. I. Stem and leaf," *Jour. Arnold Arb.*, **29**: 77–89.

———, ———, and SMITH, A. C., 1943. "The family Himantandraceae," *Jour. Arnold Arb.*, **24**: 190–206.

———, and SMITH, A. C., 1942. "Degeneriaceae, a new family of flowering plants from Fiji," *Jour. Arnold Arb.*, **23**: 356–365.

———, and SWAMY, B. G. L., 1948. "*Amborella trichopoda* Baill., a new morphological type of vesselless dicotyledon," *Jour. Arnold Arb.*, **29**: 245–254.

———, ———, 1949. "The morphology and relationships of *Austrobaileya*," *Jour. Arnold Arb.*, **30**: 211–226.

———, and THOMPSON, W. P., 1918. "Additional notes upon the angiosperms *Tetracentron*, *Trochodendron*, and *Drimys*," *Ann. Bot.*, **37**: 503–512.

———, and TUPPER, W. W., 1918. "Size variations in tracheary cells. I. A comparison between the secondary xylems of vascular cryptogams, gymnosperms, and angiosperms," *Proc. Amer. Acad. Arts & Sci.*, **54**: 149–204.

BARGHOORN, E. S., 1940. "The ontogenetic development and phylogenetic specialization of rays in the xylem of dicotyledons. I. The primitive ray structure," *Amer. Jour. Bot.*, **27**: 918–928.

———, 1941a. "The ontogenetic development and phylogenetic specialization of rays in the xylem of dicotyledons. II. Modification of the multiseriate and uniseriate rays," *Amer. Jour. Bot.*, **28**: 273–282.

———, 1941b. "The ontogenetic development and phylogenetic specialization of rays in the xylem of dicotyledons. III. The elimination of rays," *Bull. Torrey Bot. Club*, **68**: 317–325.

BOUREAU, E., 1954. *Anatomie végétale*, Vol. I, Paris: Presses Universitaires de France.

———, 1956. *Ibid.*, Vol. II.

Boureau, E., 1957. *Ibid.,* Vol. III.

Canright, J. E., 1955. "The comparative morphology and relationships of the Magnoliaceae. IV. Wood and nodal anatomy," *Jour. Arnold Arb.,* **36:** 119–140.

Carlquist, S., 1957. "Wood anatomy of Mutisieae (Compositae)," *Trop. Woods,* **106:** 29–45.

———, 1958. "Wood anatomy of Heliantheae (Compositae)," *Trop. Woods,* **108:** 1–30.

———, 1960a. "Wood anatomy of Cichorieae (Compositae)," *Trop. Woods,* **112:** 65–91.

———, 1960b. "Wood anatomy of Astereae (Compositae)," *Trop. Woods,* **113:** 54–84.

Chalk, L., 1933. "Multiperforate plates in vessels, with special reference to the Bignoniaceae," *Forestry,* **7:** 16–22.

———, 1937. "The phylogenetic value of certain anatomical features of dicotyledonous woods," *Ann. Bot.,* n.s., **1:** 409–428.

———, 1944. "On the taxonomic value of the anatomical structure of the vegetative organs of the dicotyledons. 2. The taxonomic value of wood anatomy," *Proc. Linn. Soc. London,* **155:** 214–218.

———, and Chattaway, M. Margaret, 1933. "Perforated ray cells," *Proc. Royal Soc. London,* B **113:** 89–92.

———, ———, 1934. "Measuring the length of vessel members," *Trop. Woods,* **40:** 19–26.

———, ———, 1937. "Identification of woods with included phloem," *Trop. Woods,* **50:** 1–31.

Chattaway, M. Margaret, 1932. "Proposed standards for numerical values used in describing woods," *Trop. Woods,* **29:** 20–29.

———, 1933. "Tile cells in the rays of the Malvales," *New Phyt.,* **32:** 261–273.

———, 1936. "Relation between fibre and cambial initial length in dicotyledonous woods," *Trop. Woods,* **46:** 16–20.

———, 1955. "Crystals in woody tissues. Part I," *Trop. Woods,* **102:** 55–74.

———, 1956. *Ibid.,* Part II, *Trop. Woods,* **104:** 100–124.

Cheadle, V. I., 1942. "The occurrence and types of vessels in the various organs of the plant in the Monocotyledoneae," *Amer. Jour. Bot.,* **29:** 441–450.

———, 1943a. "The origin and certain trends of specialization of the vessel in the Monocotyledoneae," *Amer. Jour. Bot.,* **30:** 11–17.

———, 1943b. "Vessel specialization in the late metaxylem of the various organs in the Monocotyledoneae," *Amer. Jour. Bot.,* **30:** 484–490.

———, 1944. "Specialization of vessels within the xylem of each organ in the Monocotyledoneae," *Amer. Jour. Bot.,* **31:** 81–92.

———, 1953. "Independent origin of vessels in the monocotyledons and dicotyledons," *Phytomorphology,* **3:** 23–44.

———, 1955. "The taxonomic use of specialization of vessels in the metaxylem of Gramineae, Cyperaceae, Juncaceae, and Restionaceae," *Jour. Arnold Arb.,* **36:** 141–157.

———, 1956. "Research on xylem and phloem—progress in fifty years," *Amer. Jour. Bot.,* **43:** 719–732.

CHOWDHURY, K. A., 1936. "Terminal and initial parenchyma cells in the wood of *Terminalia tomentosa* W. & A.," *New Phyt.,* **35:** 351–358.

———, 1939–1940. "The formation of growth rings in Indian trees," *Indian For. Rec.,* Util. **2**(1): 1–39; **2**(2): 40–57; **2**(3): 59–75.

Committee on Nomenclature, International Association of Wood Anatomists, 1957. "International glossary of terms used in wood anatomy," *Trop. Woods,* **107:** 1–36.

Committee on Standardization of Terms of Cell Size, International Association of Wood Anatomists, 1937. "Standard terms of length of vessel members and wood fibers," *Trop. Woods,* **51:** 21.

COZZO, D., 1953. "The structure and diagnostic significance of crateriform bordered pits in the vessels of *Cercidium,*" *Jour. Arnold Arb.,* **24:** 187–190.

FROST, F. H., 1929. "Bordered pits in parenchyma," *Bull. Torrey Bot. Club,* **56:** 259–263.

———, 1930a. "Specialization in secondary xylem in dicotyledons. I. Origin of vessel," *Bot. Gaz.,* **89:** 67–94.

———, 1930b. "Specialization in secondary xylem in dicotyledons. II. Evolution of end wall of vessel segment," *Bot. Gaz.,* **90:** 198–212.

———, 1931. "Specialization in secondary xylem in dicotyledons. III. Specialization of lateral wall of vessel segment," *Bot. Gaz.,* **91:** 88–96.

GARRATT, G. A., 1933a. "Systematic anatomy of the woods of the Myristicaceae," *Trop. Woods,* **35:** 6–48.

———, 1933b. "Bearing of wood anatomy on the relationships of the Myristicaceae," *Trop. Woods,* **36:** 20–44.

———, 1934. "Systematic anatomy of the woods of the Monimiaceae," *Trop. Woods,* **39:** 18–44.

GILBERT, S. G., 1940. "Evolutionary significance of ring porosity in woody angiosperms," *Bot. Gaz.,* **102:** 105–120.

HALL, J. W., 1952. "The comparative anatomy and phylogeny of the Betulaceae," *Bot. Gaz.,* **113:** 235–270.

HARRAR, E. S., 1946. "Notes on starch grains in septate fibre-tracheids," *Trop. Woods,* **85:** 1–9.

HESS, R. W., 1950. "Classification of wood parenchyma in dicotyledons," *Trop. Woods,* **96:** 1–20.

INGLE, H. D., and DADSWELL, H. E., 1953. "The anatomy of the timbers of the southwest Pacific area. II. Apocynaceae and Annonaceae," *Austral. Jour. Bot.,* **1:** 1–26.

———, ———, 1956. "The anatomy of the timbers of the southwest Pacific area. IV. Cunoniaceae, Davidsoniaceae, and Eucryphiaceae," *Austral. Jour. Bot.,* **4:** 125–151.

KRIBS, D. A., 1935. "Salient lines of structural specialization in the wood rays of dicotyledons," *Bot. Gaz.,* **96:** 547–557.

———, 1937. "Salient lines of structural specialization in the wood parenchyma of dicotyledons," *Bull. Torrey Bot. Club,* **64:** 177–186.

METCALFE, C. R., and CHALK, L., 1950. *Anatomy of the dicotyledons,* Oxford: Clarendon Press.

MOSELEY, F. M., 1948. "Comparative anatomy and phylogeny of the Casuarinaceae," *Bot. Gaz.*, **110**: 231–280.

————, and BEEKS, R. M., 1955. "Studies on the Garryaceae. I. The comparative morphology and phylogeny," *Phytomorphology*, **5**: 314–346.

MOSS, E. H., 1940. "Interxylary cork in *Artemisia* with a reference to its taxonomic significance," *Amer. Jour. Bot.*, **27**: 762–768.

NAST, CHARLOTTE G., and BAILEY, I. W., 1946. "Morphology of *Euptelea* and comparison with *Trochodendron*," *Jour. Arnold Arb.*, **27**: 186–192.

RECORD, S. J., 1934. *Identification of the timbers of temperate North America,* New York: John Wiley.

————, and CHATTAWAY, M. MARGARET, 1939. "List of anatomical features used in classifying dicotyledonous woods," *Trop. Woods*, **57**: 11–16.

RENDLE, B. J., 1937. "Gelatinous wood fibers," *Trop. Woods*, **52**: 11–19.

————, and CLARKE, S. H., 1934a. "The problem of variation in the structure of wood," *Trop. Woods*, **38**: 1–8.

————, ————, 1934b. "The diagnostic value of measurements in wood anatomy," *Trop. Woods*, **40**: 27–37.

RODRIGUEZ, R. L., 1957. "Systematic anatomical studies on *Myrrhidendron* and other woody Umbellales," *Univ. Calif. Publ. Bot.*, **29**: 145–318.

SHUTTS, G. F., 1960. "The wood anatomy of the Hernandiaceae and Gyrocarpaceae," *Trop. Woods*, **113**: 85–123.

SOLEREDER, H., 1908. *Systematic anatomy of the dicotyledons* (trans. by Boodle and Fritsch), Oxford: Clarendon Press.

STERN, W. L., 1952. "The comparative anatomy of the xylem and the phylogeny of the Julianiaceae," *Amer. Jour. Bot.*, **39**: 220–229.

————, 1954a. "Comparative anatomy of xylem and phylogeny of Lauraceae," *Trop. Woods*, **100**: 1–72.

————, 1954b. "A suggested classification for intercellular spaces," *Bull. Torrey Bot. Club*, **81**: 234–235.

————, 1955. "Xylem anatomy and relationships of Gomortegaceae," *Amer. Jour. Bot.*, **42**: 874–885.

————, and BRIZICKY, G. K., 1958. "The comparative anatomy and taxonomy of *Heteropyxis*," *Bull. Torrey Bot. Club*, **85**: 111–123.

————, and GREENE, S., 1958. "Some aspects of variation in wood," *Trop. Woods*, **108**: 65–71.

SWAMY, B. G. L., and BAILEY, I. W., 1949. "The morphology and relationships of *Cercidiphyllum*," *Jour. Arnold Arb.*, **30**: 187–210.

————, ————, 1950. "*Sarcandra,* a vesselless genus of the Chloranthaceae," *Jour. Arnold Arb.*, **31**: 117–129.

TAKHTAJAN, A. L., 1959. *Essays on the evolutionary morphology of plants* (trans. by O. H. Gankin), Washington, D. C.: American Institute of Biological Sciences.

THOMPSON, W. P., 1923. "The relationships of the different types of angiospermic vessels," *Ann. Bot.*, **37**: 183–192.

TIPPO, O., 1938. "Comparative anatomy of the Moraceae and their presumed allies," *Bot. Gaz.*, **100**: 1–99.

TIPPO, O., 1940. "The comparative anatomy of the secondary xylem and the phylogeny of the Eucommiaceae," *Amer. Jour. Bot.*, **27:** 832–838.

———, 1941. "A list of diagnostic characteristics for descriptions of dicotyledonous woods," *Trans. Illinois Acad. Sci.*, **34:** 105–106.

———, 1946. "The role of wood anatomy in phylogeny," *Amer. Midl. Nat.*, **36:** 362–372.

TUPPER, W. W., 1927. "Woods with conspicuously large rays," *Trop. Woods,* **11:** 5–9.

VANDER WYK, R. W., and CANRIGHT, J. E., 1956. "The anatomy and relationships of the Annonaceae," *Trop. Woods*, **104:** 1–24.

VESTAL, P. A., 1940. "Wood anatomy as an aid to classification and phylogeny," *Chronica Botanica*, **6:** 53–54.

WARDROP, A. B., and DADSWELL, H. E., 1955. "The nature of reaction wood. IV. Variations in cell wall organization of tension wood fibers," *Austral. Jour. Bot.*, **3:** 177–189.

WILSON, T. K., 1960. "The comparative morphology of the Canellaceae. I. Synopsis of genera and wood anatomy," *Trop. Woods*, **112:** 1–27.

WOODWORTH, R. H., 1932. "Diaxylary laticiferous cells of *Beaumontia grandiflora*," *Jour. Arnold Arb.*, **13:** 35–36.

———, 1935. "Fibriform vessel members in the Passifloraceae," *Trop. Woods*, **41:** 8–16.

chapter five ❯ Phloem

Because xylem, particularly secondary xylem, has provided a gamut of material interpretable for evolutionary and taxonomic purposes, various investigators have devoted attention to phloem to determine if similar phylogenetically significant trends occur there. Xylem and phloem, however, incorporate different characteristics and different cell types. The two tissues require different means of analysis as well as preparation. Phloem consists of sieve cells or sieve-tube elements and companion cells, parenchyma, and often fibers. Phloem fibers are discussed in Chapter 8. Sieve cells are considered a primitive type of phloem conductive element. They are not arranged in axial files and their end walls are not different in structure from lateral walls. Sieve-tube elements have end walls differentiated into sieve plates, which are more specialized than the sieve areas found on lateral walls of sieve-tube elements and on all walls of sieve cells. Sieve-tube elements are arranged in axial files and are associated with companion cells, which are sister cells of sieve-tube elements ontogenetically. The term "sieve element" applies both to sieve cell and to sieve-tube element.

Primitive sieve elements

Gymnosperms and vascular cryptogams have been used in formulation of concepts of primitive sieve elements, just as they have been used in demonstrating primitive types of tracheary tissue (Esau *et al.*, 1953). Primitively, phloem contains sieve cells. Such cells have been reported among angiosperms only in *Austrobaileya scandens* (Bailey and Swamy, 1949) and *Sorbus aucuparia* of

Rosaceae (Huber, 1939). Because of the known trends toward shortening of cambial initials in dicots, long sieve elements in secondary phloem may be considered primitive in the same way as long vessel elements. Such measurements, however, must take into account the fact that unlike vessel elements, sieve-tube elements are frequently derived from cambial initials following transverse septation of the initial (Cheadle and Esau, 1958; Zahur, 1959). Thus, short sieve-tube elements do not necessarily indicate an advanced condition, and studies on length must note whether this septation takes place.

Hemenway (1911, 1913), Huber (1939), Cheadle and Whitford (1941), and Cheadle (1948) agree, on the basis of various studies, that a long end wall, with numerous sieve areas, is primitive in sieve-tube elements, and Zahur (1959) shows that these features, like end-wall characteristics of vessel elements, show statistical correlation. This correlation does not appear to be merely a morphogenetic one, because sieve areas on an end wall of a given length may be quite various in morphology. Zahur (1959) regards small pores in the sieve plate as primitive.

Evolution in sieve-tube elements

As early as 1913, phloem of vascular plants was surveyed by Hemenway in an attempt to demonstrate trends of evolution. Hemenway concluded that lateral sieve areas become smaller and less conspicuous, while the terminal area becomes clearly differentiated as a sieve plate. Moreover, he found that the sieve plate tends to change in position from oblique to nearly horizontal, and from compound in structure to simple, during phylogenetic advance. These conclusions are like those reached later by Huber (1939). Working on a statistical basis with metaphloem of monocots, Cheadle and his co-worker validated and expanded these conclusions. They found, for example, a high degree of statistical correlation between simple sieve plates and plates oriented transversely or nearly so (Cheadle and Whitford, 1941; Cheadle, 1948). Such correlations are also reported between compound sieve plates and those that are very oblique. Zahur (1959), in a rather extensive survey of dicot phloems, has attempted to develop criteria for indicating phylogenetic advance. Despite the difficulties posed by septation of sieve-tube element initials, Zahur does find a correlation between length of sieve tube and length of sieve plate. Zahur utilizes three categories (originated by Hemenway) for his purposes:

I. Sieve-tube elements long, with very oblique sieve plates containing 10 or more sieve areas (if sieve-area number is unreliable, the other criteria are used).

II. Sieve-tube elements of medium length, with oblique sieve plates containing 2 to 10 sieve areas.

III. Sieve-tube elements short, with slightly oblique to transverse, simple plates.

Zahur compares these types to other features in a statistical fashion. His conclusions are as follows: (1) Sieve-tube elements have been shortened in the course of evolution, both by decrease in length of initials and by septation of initials. Zahur's suggestion that degree of secondary septation is an indication of phylogenetic advance is open to question, however. To be sure, little if any secondary septation of cambial initials takes place in gymnosperms. However, the families he lists as having considerable septation (for example, Monimiaceae) are by no means advanced in many other respects, and his statement that some families that have considerable transverse septation are woody groups derived from herbs is certainly not likely in all cases—for example, Saxifragaceae (Hydrangeaceae, of some authors, for Zahur's species), which has remarkably primitive wood for dicots. Zahur expresses the tendency toward septation of initials in angiosperms as a "retention of meristematic activity" that has been formed in angiosperms *de novo*. Next (2) Zahur concludes (in agreement with earlier workers) that there has been a phylogenetic decrease in length of end walls and number of sieve areas. In my opinion, this may be true, but one must explain why families such as Compositae, which have only simple vessel perforation plates (or some special form of plate), frequently have compound sieve plates on rather oblique end walls. Either sieve-tube elements show a marked lag in evolution compared with vessel elements (in a given plant), or the sequences indicated for sieve elements are to some extent reversible. That the latter is possible may be indicated by the fact that subdivision or partial subdivision of sieve areas is frequent, whereas an analogous phenomenon is probably not present in vessel elements. Moreover, vessel elements show a clearly marked evolution in lateral wall pitting, whereas, as

Zahur admits, evolution of sieve areas on lateral walls is by no means clear (or necessarily unidirectional). The trend from clearly organized sieve areas on lateral walls to vague or indefinite ("poorly organized") sieve areas that was postulated by Hemenway (1913) has been questioned by MacDaniels (1918) and Zahur (1959) because of the many exceptions to this trend they found. Thus, nearly half of the species classified by Zahur as having Type I sieve-tube elements had indefinite sieve areas on lateral walls. (3) Zahur also concludes that size of pores in sieve plates of dicot secondary phloem has increased. And finally, (4) Zahur claims that there is no overall phylogenetic trend in presence or absence of nucleoli in functional sieve tubes, although this may be a feature of taxonomic importance in some groups.

Evolution in phloem as a tissue

Trends in sieve-tube-element evolution may be compared with varied expressions in other features of phloem. Just as Cheadle established trends of evolution in tracheary tissue of monocots (see Chapter 4), Cheadle and Whitford (1941) established a sequence of specialization in metaphloem of monocots. Interestingly, the trends of specialization appear to be the reverse of xylem trends, for the least advanced sieve elements occur in monocot roots and the most advanced in leaves, inflorescence axes, corms, and rhizomes, whereas those of aerial stems are intermediate. In an entirely different connection, Cheadle and Uhl (1948) used types of bundle construction in monocots to establish trends of phloem conformation. These bundle types (see Chapter 8) had been given phylogenetic ratings by means of rather clear statistical correlations with vessel types. They found that the "regular" type of arrangement of sieve tubes, such as may be seen in Gramineae, is correlated with more advanced bundle types, whereas a lack of specificity in orientation of companion cells to sieve-tube elements occurs in more primitive types of bundle construction. They do not find that sieve tube element-companion cell arrangement has evolved at the same rate as vessel types, when compared to the bundle types, and they explain this by suggesting that metaphloem shows an evolutionary lag compared with metaxylem.

Various trends in evolution of secondary phloem of dicots are suggested by Zahur (1959). He distinguishes three types of companion cells: (1) those that occur singly and are much shorter than the sieve-tube elements; (2) those that are about as long as the sieve-tube element they accompany; and (3) those that are as long as the sieve-tube element but are septate. Each of these types he finds characteristic of natural groupings. He finds that the first type may be considered more primitive, the latter two types advanced. Cheadle and Esau (1958) suggest that decrease in number of companion cells may mark evolutionary advance. Likewise, shortening of the sieve-tube element may be correlated with increasing contact between sieve-tube element and companion cells, a change that is achieved by altered arrangement between the two types of cells. Resch (1954) has even suggested, on the basis of primary phloem of *Vicia faba*, that there may be various degrees of specialization among companion cells in a single stem.

Other features of secondary phloem that Zahur finds may be of comparative significance, although they may not be related to major trends of phloem evolution, include distribution (irregular vs. banded) of phloem parenchyma, morphology of phloem parenchyma cells, and the nature of sclerenchyma in secondary phloem. Unlike libriform fibers in secondary xylem, which represent phylogenetic derivatives of tracheids, fibers in secondary phloem bear no such phylogenetic relation to sieve elements. Because of this fact, and because periderm, in addition to secondary phloem, is involved in bark formation, information on stem sclerenchyma and periderm are considered in Chapter 8.

Conclusions

One may conclude from the comparative data of Zahur and the papers of other workers that phloem contains a number of systematically usable features. The major trends of phloem evolution are by no means similar in all points to the major trends of xylem evolution. Evolutionary trends in phloem seem much more difficult to establish clearly, for reasons suggested above. Nevertheless, the careful work of Cheadle and Esau (1958) on Calycanthaceae suggests that if sufficient caution is used, valid comparative data may be obtained. We cannot yet state that certain trends in phloem evolution are irreversible, and exceptional cases provide difficulties in interpretation. Phloem probably offers a much more limited range of features than does xylem, and different means of preparation and study must be used. Nevertheless, phloem cannot be overlooked as a source of comparative material, and we may look forward to its increasing usefulness as it becomes better understood.

REFERENCES

BAILEY, I. W., and SWAMY, B. G. L., 1949. "The morphology and relationships of *Austrobaileya*," *Jour. Arnold Arb.*, **30:** 211–226.

CHEADLE, V., 1948. "Observations on the phloem in the Monocotyledoneae. II. Additional data on the occurrence and phylogenetic specialization in structure of the sieve tubes in the metaphloem," *Amer. Jour. Bot.*, **35:** 129–131.

———, and ESAU, KATHERINE, 1958. "Secondary phloem of Calycanthaceae," *Univ. Calif. Publ. Bot.*, **29:** 397–510.

———, and UHL, NATALIE W., 1948. "The relation of metaphloem to the types of vascular bundles in the Monocotyledoneae," *Amer. Jour. Bot.*, **35:** 578–583.

———, and WHITFORD, NATALIE B., 1941. "Observations on the phloem in the Monocotyledoneae. I. The occurrence and phylogenetic specialization in structure of the sieve tubes in the metaphloem," *Amer. Jour. Bot.*, **28:** 623–627.

ESAU, KATHERINE, CHEADLE, V., and GIFFORD, E., 1953. "Comparative structure and possible trends of specialization of the phloem," *Amer. Jour. Bot.*, **40:** 9–19.

HEMENWAY, A. F., 1911. "Studies on the phloem of the dicotyledons. I. Phloem of the Juglandaceae," *Bot. Gaz.*, **51:** 131–135.

———, 1913. "Studies on the phloem of the dicotyledons. II. The evolution of the sieve-tube," *Bot. Gaz.*, **55:** 236–243.

HUBER, B., 1939. "Siebröhrensystem unserer Bäume und seine jahreszeitlichen Veränderungen," *Jahrb. Wiss. Bot.*, **88:** 176–242.

MACDANIELS, L. H., 1918. "The histology of the phloem in certain woody angiosperms," *Amer. Jour. Bot.*, **5:** 347–378.

RESCH, A., 1954. "Beiträge zur Cytologie des Phloems. Entwicklungsgeschichte der Siebröhrenglieder und Geleitzellen bei *Vicia faba* L.," *Planta*, **44:** 75–98.

ZAHUR, M. Z., 1959. "Comparative study of secondary phloem of 423 species of woody dicotyledons belonging to 85 families," *Cornell Univ. Agr. Stat. Mem.*, **358:** 1–160.

chapter six

Vascularization
of the Shoot

With the advent of the stelar concept, the vascular system of the plant came to be regarded as a three-dimensional structure best studied as such rather than as isolated transections of, for example, the petiole alone. This broader concept is utilized in this chapter, which is concerned with vascular patterns other than those of the root and flower.

SEEDLING

Comparative studies. A vast literature has accumulated in this field. With respect to gross morphology and some aspects of anatomy, the volumes of Lubbock (1892) remain an important compendium. Methods for representing or describing seedling anatomy often leave room for improvement, and in this respect, the rather comprehensive type of illustrations used by Gravis (1943) are commendable. Several of these, shown as Fig. 6-1, offer ease of interpretation. In dicots, many studies have been arranged by family or order groupings; studies by Compton (1922), Davey (1916), de Fraine (1910), Hill and de Fraine (1912), Lee (1912, 1914), and Thomas (1914) are examples. These, as well as papers cited by Boureau (1954), include much of the literature in this field. For a survey of seedling anatomy in dicots at large, see Tronchet (1930) and Gravis (1943).

In monocots, Lucy Boyd (1932) has compiled much comparative data. For analysis of the complexities of palm seedlings, the interesting work of Gineis (1952) is worthy of note.

Analysis of seedling structure should involve sampling techniques, because as noted by Boureau (1954), variations within species occur. Such variations have been ingeniously plotted as

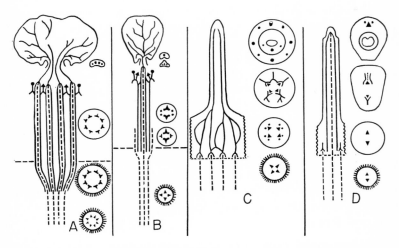

Fig. 6-1. Conventionalized representations of vascular anatomy of seedlings of two dicots (left) and two monocots (right). These conventions permit a three-dimensional picture of seedling vascularization. In the dicots, half of the seedling (left) is diagramed; in the monocots, the cylindrical structure of the seedling is shown laid flat; in all, transections of four levels, representing root, hypocotyl, transition region, and cotyledon, are given. Broken lines and pointed symbols = exarch xylem; unbroken lines and rounded symbols = endarch xylem. A, *Fagus sylvatica*; B, *Carpinus betulus*; C, *Ornithogalum caudatum*; D, *Allium cepa*. In each pair, the vascular pattern of the seedling at right suggests reduction. (Redrawn from Gravis, 1943.)

curves for certain species by Gineis (1952), so that modal conditions and extremes are clearly visible.

Evolution. Seedling anatomy has been used as a basis for discussion of origin of monocots from dicots and the manner in which such origin could have taken place. For reviews on this topic, see Arber (1925) and Hill (1906). Vascular anatomy of cotyledons is perhaps not of much significance in this case, because the duality of traces in each dicot cotyledon is as frequent as the well-known duality of traces in monocot cotyledons.

The nature of the transition region has naturally played an important part in morphological interpretations of angiosperm structure, as the papers of Chauveaud (1910) and Compton (1912) show. Complexity of vascularization by no means parallels size of seedling (Boureau, 1954), and the conservatism of seedling anatomy undoubtedly does offer material for studies on evolutionary tendencies (Lenoir, 1920). Several workers (Thomas, 1914; Tronchet, 1930) have suggested a transition from tetrarch to diarch seedling structure. Vascularization in seedlings may well have decreased in complexity (for example, the right seedling in each pair, Fig. 6-1). Increase in seedling complexity has also doubtless occurred, as seems to be the case in seedlings that possess multilacunar nodes, such as *Gustavia* (Lignier, 1890) and *Fitchia* (Carlquist, 1957).

The nature of the cotyledonary node offers much material to development of theories of nodal anatomy, and such materials are considered below in this connection.

NODAL ANATOMY

Major nodal types. Because monocot nodes are highly complex structures markedly different from most dicot nodes, terminology and studies of nodal anatomy apply to dicot nodes chiefly. Currently, four main types of dicot nodes are recognized (Fig. 6-2A-D): (A) unilacunar, two-trace, in which the two traces are connected to opposite halves of the eustele (see also Fig. 6-3A); (B) unilacunar, with a single trace; (C) trilacunar, with three traces from three gaps; and (D) multilacunar, in which more than three traces and three gaps per node are present. In an effort to explore the distribution and significance of nodal types, Sinnott and Bailey (1914a) compiled a great deal of data on the occurrence of unilacunar (not further differentiated), trilacunar, and multilacunar types. According to this study, the trilacunar is widespread and seems to be a central type from which, in various groups, multilacunar (Fig. 6-2, sequence C to D) and unilacunar (Fig. 6-2, sequence C to B) have originated. Another type of derivation seemingly occurs in Ericales. In this order, which is predominantly unilacunar, multilacunar types (Fig. 6-2, sequence B to D) have originated in the Epacridaceae.

The discovery of the fourth type of nodal anatomy (unilacunar, two-trace) by Marsden and Bailey (1955) led to revision of concepts of nodal evolution. The trilacunar type remains a form from which the unilacunar (one-trace) and multilacunar types have been derived, but the unilacunar two-trace type, as shown in Figs. 6-2A and 6-3A, E, now appears basic in angiosperms. Reasons for this are given by Marsden and Bailey (1955), Bailey (1956), and Canright

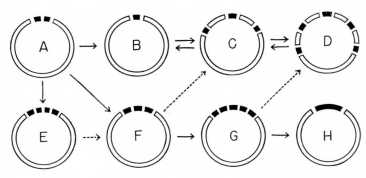

Fig. 6-2. Evolutionary patterns of nodal anatomy in dicots. A, unilacunar, two-trace; B, unilacunar, one-trace; C, trilacunar; D, multilacunar; E, unilacunar, four-trace; F, unilacunar, three-trace; G, unilacunar, multiple trace; H, unilacunar, compound single trace. Traces are shown in black. Further explanation in text. (Modified from Canright, 1955.)

(1955), and may be summarized as follows: (1) In vascular plants other than angiosperms, the unilacunar two-trace condition seems to be widespread and basic. If angiosperms have a "pteropsid" ancestry, this would be expected. (2) Studies of the cotyledonary node reveal that of dicots studied, 77 percent have an even number of traces (Bailey, 1956). The cotyledonary node appears conservative (compared with other nodes) in retaining this condition. Just as many plants with multilacunar nodes in the adult plant have trilacunar nodes in juvenile leaves of seedlings, the nodal type in cotyledons appears less alterable than that of later formed leaves. This is confirmed by transitional conditions that may be found in seedlings (Bailey, 1956). (3) Dicots with many other primitive features show a unilacunar two-trace condition or some derivative of it. For example, *Austrobaileya* (Fig. 6-3E) shows this nodal type throughout the entire plant, including the flower (Fig. 10-1A). Chloranthaceae show this condition in *Ascarina* (Fig. 6-3A) and a series of derivative conditions in *Chloranthus* (Fig. 6-3C), *Sarcandra* (Fig. 6-3D), and *Hedyosmum* (Fig. 6-3B). This sequence is represented in Fig. 6-2 by the succession A-F-G. Alterations of the unilacunar two-trace condition in Monimiaceae are illustrated in Fig. 6-2 by the sequence A-E-F-G-H, which represent *Trimenia*, *Piptocalyx*, *Hortonia*, *Mollinedia*, and *Atherosperma* respectively (Money, Bailey, and Swamy, 1950). Other families in which the two-trace unilacunar condition (or some modification of it) occurs include Trimeniaceae (Money, Bailey, and Swamy, 1950), Amborellaceae (Bailey and Swamy, 1948), Calycanthaceae (Fahn and Bailey, 1957), Lactoridaceae (Bailey, 1956), Gomortegaceae, Lauraceae, and Hernandiaceae.

The occurrence of the unilacunar two-trace condition in such an advanced family as Verbenaceae (Marsden and Bailey, 1955) is curious, however. New surveys are needed to establish phylogenetic significance of nodal types in particular groups.

Taxonomic usefulness. As a systematic feature, nodal anatomy is best used in combination with petiolar vascularization, as emphasized below. Comparisons of nodes may be valuable for showing relationship or distinctness of genera or even species (Sinnott and Bailey, 1914a). In making such distinctions it is convenient to pair type of node with type of phyllotaxy (that is, opposite leaves, multilacunar node) (Carlquist, 1959). Examples of systematic applications of nodal data are offered by the work of Bailey and Howard (1941) and Canright (1955). Note should be taken of the fact that within a single plant, nodal anatomy can vary (Bailey, 1956). This fact has been graphically portrayed by Post (1958), demonstrating that sampling may be necessary for determinations of nodal conditions.

Nodal anatomy and stipules. Sinnott and Bailey (1914b) showed that a relation exists between trilacunar or multilacunar nodes and possession of stipules; unilacunar nodes are exstipulate. These relationships have subsequently been confirmed (Dormer, 1944; Bailey, 1956). Phylogenetic speculation as to significance of stipules does not seem warranted as yet. Venation of stipules does

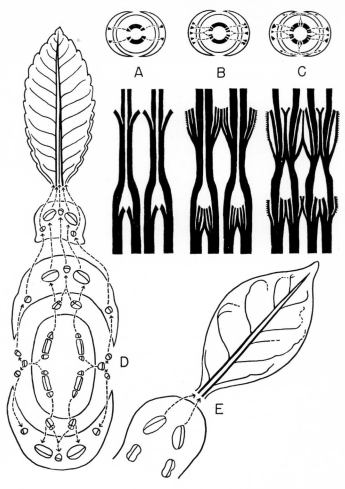

Fig. 6-3. Representations of primitive types of nodal anatomy in Chloranthaceae (A-D) and Austrobaileyaceae (E). In A-C, a transection and a laid-out vascular cylinder with two successive nodes are shown for each species. Broken lines represent course of bundles. In vascular cylinders for B and C, the toothed edges of bundles indicate that they are halves of a single bundle (other half at opposite edge). D shows node-petiole anatomy as a single unit, with broken lines indicating course of bundles. A, *Ascarina;* B, *Hedyosmum;* C, *Chloranthus;* D, *Sarcandra;* E, *Austrobaileya.* (Redrawn: A-C from Swamy, 1953; D from Swamy and Bailey, 1950; E from Bailey and Swamy, 1949.)

not differ appreciably from that of basal leaflets of a compound leaf or the basal portion of a simple leaf. Therefore, there seems no reason to believe that stipules have not been derived from exstipulate leaves, or, on the contrary, that stipulate leaves have not given rise to exstipulate ones (Howard, unpublished).

Stem vascularization. Careful studies on longitudinal course of bundles in the stem have been made in recent years. These are usually illustrated in an "unrolled cylinder" manner, as shown in Fig. 6-3A-C. As more of these patterns become available, the likelihood of their use in systematics will increase. Dormer (1945, 1946) has utilized some of these patterns, and claims an "open system" and "insertions in contact" to be primitive in Leguminosae, with opposing conditions advanced. In a different manner, Fahn and Arzee (1959) have utilized number and path of stem bundles as characteristics within Chenopodiaceae. Cortical and medullary bundles are considered below under the heading "Additional Vascularization Patterns."

NODE-PETIOLE ANATOMY

With our increasing knowledge of both nodes and petioles, separation of one from the other becomes increasingly artificial. To be sure, comparative data can, and have been, developed on the basis of a single petiolar section for each taxon. However, Howard (1959) has wisely recommended methods that utilize sections representing levels below a node, through a node, through the petiole (enough levels to secure representation of the "vascular events"), and into the midrib of the leaf. Thus a complete gamut of vascular patterning is revealed. Such a method is exhibited by Fig. 6-3D. More complicated petioles may require more sections than simpler ones. Significantly, ontogenetic studies suggest that there is no single point along a petiole that could be called "characteristic" or "typical," or be used to the exclusion of other sections for comparative purposes. Howard (1959) has suggested a system of classification based first upon nodal types, then within these groups upon distinctive configurations of the petiolar stele. These types are based upon xylem alone; phloem may, in addition, be present as isolated strands. A fragment of this system may be illustrated as follows:

"3–3" (three traces from three gaps):

> Traces form an arc, the edges of which inroll, forming, by separation from the arc, medullary bundles (*Quercus, Tilia*).
>
> Traces form two concentric circles (*Fagus*).
>
> Traces form into a polystelic condition, often with an I-shaped configuration in a flattened portion of the petiole (*Platanus, Populus*).
>
> Etc.

This type of classification, which will be presented in a forthcoming supplement to Metcalfe and Chalk's *Anatomy of the Dicotyledons,* will prove to be a useful one. Values of Howard's method lie both in providing a comprehensive picture of node-petiole patterns within a single plant and in providing a method that can be applied universally throughout dicots.

In the existing literature on petiolar anatomy, one finds a variety of methods and criteria employed, depending on the taxonomic group involved. Metcalfe and Chalk (1950) should be consulted for summaries of such literature in particular families. Recent studies that show the systematic significance of petiolar anatomy include those of Comyn (1957) on Rubiaceae, Dehay on Euphorbiaceae (1935) and Sterculiaceae (1941), and Watari on Leguminosae (1934) and *Acer* (1936). A résumé of systematic usefulness of petiolar anatomy is given by Hare (1944).

Vascular anatomy of the petiole does not seem usable as a criterion for distinguishing herbaceous from woody plants. Although many herbs do have traces that extend from the node into the lamina without appreciable branching, grouping, or rearrangement, this is also true of many woody plants, such as Winteraceae (Bailey and Nast, 1944) and Pittosporaceae.

ADDITIONAL VASCULARIZATION PATTERNS

The problems of "cortical" and "medullary" bundles have often been studied independently of nodal conditions, and the two problems may seem unrelated to the reader. When both problems are studied in a three-dimensional manner rather than from arbitrary (usually internodal) transections, however, one finds that medullary or cortical bundles very frequently are—at least in part—related in their course to nodal "events." This was clearly seen by Col (1902), who claimed that "almost all medullary bundles are abnormal continuations of normal bundles, and their relative position alone constitutes the anomaly." Few cases of cortical or medullary

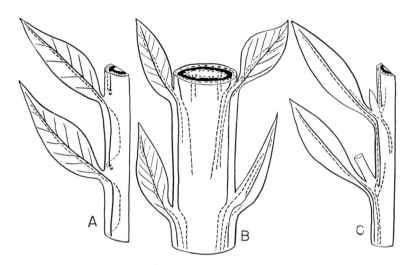

Fig. 6-4. Three-dimensional representations of dicot stems with pith bundles, showing how these are related to nodal conditions. Broken lines represent pith bundles and their upward extension, if any, into leaves; unbroken lines (and black circle at top of stem) indicate ordinary bundles of the vascular cylinder. A, *Phytolacca* sp., showing that medullary bundles in this species are merely inwardly displaced bundles of the cylinder; B, *Campanula rapunculoides*, an instance in which pith bundles are freely terminating downward extensions of leaf traces and bundles of the cylinder; C, *Symphyandra pendula*, in which pith bundles terminate freely in the pith in both acropetal and basipetal directions. (Redrawn from Col, 1904.)

bundles show isolated or freely terminating bundles in internodes. Because of the strong relation between nodal behavior and abnormal position of bundles in pith and cortex (Fig. 6-4), there seems more value in considering the two phenomena together rather than separately.

Recent studies of cortical bundles are admirable in this respect. For example, the study of Johnson and Truscott (1956) on *Serjania* has shown that a leaf trace may enter the cortex many nodes below the leaf supplied by this trace. Papers by Fahn and Bailey (1957) on Calycanthaceae and Fahn and Arzee (1959) on Chenopodiaceae detail the nature of connections between a network of cortical bundles and the vascular cylinder or leaf traces. Majumdar (1941) illustrates a case in which cortical bundles are downward extensions of lateral leaf traces and Lignier (1887) has described the nature of cortical bundles in Melastomaceae. At nodes of *Euptelea* (Nast and Bailey, 1946), inflorescence bundles are intermingled with leaf traces. These instances all may represent special cases, but they have obvious value to the taxonomic system.

With respect to medullary bundles, the monograph of Col (1904) seems a model in the completeness with which the three-dimensional pattern of bundles has been analyzed. Three of his figures, shown as Fig. 6-4, illustrate the relation between nodal conditions and pith bundles. Solereder (1908) and Metcalfe and Chalk (1950) list families in which medullary and cortical bundles occur; for literature in this field, the reader can consult papers cited under these families.

With regard to evolution of these anomalous types, the suggestion of Worsdell (1919) that medullary bundles represent vestiges of an ancestral scattered-bundle system does not seem to have much merit, at least as a generalization. The concept of Wilson (1924) that medullary bundles represent bundles and traces, the medullary placement of which has been gradually exaggerated by evolution, seems more plausible but leaves much room for investigation.

An extreme vascular anomaly is formed by the "polystely" of such genera as *Gunnera* (Van Tieghem and Douliot, 1886). For listing of families with this condition, see Solereder (1908).

The phenomenon of bicollateral bundles and the occurrence of "internal" phloem are in part related to the problem of medullary bundles. Comparative data on these occurrences are supplied by Baranetzky (1900) and listings of families containing these features are given by Solereder (1908) and Metcalfe and Chalk (1950). A form of vascular anomaly that is probably mostly unrelated to nodal conditions but that deserves three-dimensional study is the occurrence of strands of phloem without xylem, or xylem without phloem. Instances of these are given by Van Tieghem (1891).

There is no reason why patterns of shoot vascularization in

monocots should not yield a wealth of comparative data. Unfortunately, the high degree of complexity of monocot stems has discouraged such work in this group, but we may look forward to appearance of studies on monocot nodes as sophistication of plant anatomists on nodal matters increases.

REFERENCES

ARBER, AGNES, 1925. *Monocotyledons,* Cambridge: Cambridge University Press.

BAILEY, I. W., 1956. "Nodal anatomy in retrospect," *Jour. Arnold Arb.,* **37:** 269–287.

———, and HOWARD, R. A., 1941. "The comparative morphology of the Icacinaceae. I. Anatomy of the node and internode," *Jour. Arnold Arb.,* **22:** 125–132.

———, and NAST, CHARLOTTE G., 1944. "The comparative anatomy of the Winteraceae. IV. Anatomy of the node and vascularization of the leaf," *Jour. Arnold Arb.,* **25:** 215–221.

———, and SWAMY, B. G. L., 1948. "The morphology and relationships of *Austrobaileya,*" *Jour. Arnold Arb.,* **30:** 211–226.

———, ———, 1956. "*Amborella trichopoda* Baill., a new morphological type of vesselless dicotyledon," *Jour. Arnold Arb.,* **29:** 245–254.

BARANETZKY, J., 1900. "Recherches sur les faisceaux bicollatéraux," *Ann. Sci. Nat. Bot.,* sér. 8, **12:** 261–332.

BOUREAU, E., 1954. *Anatomie végétale,* Vol. I, Paris: Presses Universitaires de France.

BOYD, LUCY, 1932. "Monocotylous seedlings," *Trans. Bot. Soc. Edinburgh,* **31:** 1–224.

CANRIGHT, J., 1955. "The comparative morphology and relationships of the Magnoliaceae. IV. Wood and nodal anatomy," *Jour. Arnold Arb.,* **36:** 119–140.

CARLQUIST, S., 1957. "The genus *Fitchia* (Compositae)," *Univ. Calif. Publ. Bot.,* **29:** 1–144.

———, 1959. "Studies on Madinae: anatomy, cytology, and evolutionary relationships," *Aliso,* **4:** 171–236.

CHAUVEAUD, G., 1910. "Recherches sur les tissus transitoires du corps végétif des plantules vasculaires," *Ann. Sci. Nat. Bot.,* sér. 9, **12:** 1–70.

COL, A., 1902. "Sur les relations des faisceaux medullaires et des faisceaux dits surnuméraires avec les faisceaux normaux," *Jour. de Bot.,* **16:** 234–255.

———, 1904. "Recherches sur la disposition des faisceaux dans la tige et les feuilles de quelques dicotylédones," *Ann. Sci. Nat. Bot.,* sér. 8, **20:** 1–288.

COMPTON, R. H., 1912. "Theories of the anatomical transition from root to stem," *New Phyt.,* **11:** 13–25.

———, 1922. "An investigation of the seedling structure of the Leguminosae," *Jour. Linn. Soc. London,* **41:** 1–122.

Comyn, J., 1957. "Principaux aspects de l'appareil libéro-ligneux foliaire des Rubiacées," *Ann. Sci. Nat. Bot.*, sér. 11, **18:** 27–70.

Davey, A. J., 1916. "Seedling anatomy of certain Amentiferae," *Ann. Bot.*, **30:** 575–599.

de Fraine, E., 1910. "The seedling structure of certain Cactaceae," *Ann. Bot.*, **24:** 125–175.

Dehay, C., 1935. "L'appareil libéro-ligneux foliaire des Euphorbiacées," *Ann. Sci. Nat. Bot.*, sér. 10, **17:** 147–296.

———, 1941. "L'appareil libéro-ligneux foliaire des Sterculiacées," *Ann. Sci. Nat. Bot.*, sér. 11, **2:** 45–131.

Dormer, K. J., 1944. "Some examples of correlation between stipules and lateral leaf traces," *New Phyt.*, **43:** 151–153.

———, 1945. "An investigation of the taxonomic value of shoot structure in angiosperms with especial reference to Leguminosae," *Ann. Bot.*, n.s., **9:** 141–153.

———, 1946. "Vegetative morphology as a guide to the classification of the Papilionatae," *New Phyt.*, **45:** 145–161.

Fahn, A., and Arzee, Tova, 1959. "Vascularization of articulated Chenopodiaceae and the nature of their fleshy cortex," *Amer. Jour. Bot.*, **46:** 330–338.

———, and Bailey, I. W., 1957. "The nodal anatomy and the primary vascular cylinder of the Calycanthaceae," *Jour. Arnold Arb.*, **38:** 107–117.

Gineis, C., 1952. "Contribution à l'étude anatomique des plantules de palmiers. III," *Bull. Mus. Nat. Hist.*, Paris, **24:** 100–107.

Gravis, A., 1943. "Observations anatomiques sur les embryons et les plantules," *Lejeunia*, **3:** 180–196.

Hare, C. L., 1944. "On the taxonomic value of the anatomical structure of the vegetative organs of the dicotyledons. 5. The anatomy of the petiole and its taxonomic value," *Proc. Linn. Soc. London*, **155:** 223–229.

Hill, A. W., 1906. "The morphology and seedling structure of the geophilous species of *Peperomia*, together with some views on the origin of monocotyledons," *Ann. Bot.*, **22:** 395–427.

Hill, T. G., and de Fraine, E., 1912. "On the seedling structure of certain Centrospermae," *Ann. Bot.*, **26:** 175–199.

Howard, R. A., 1959. "The vascular anatomy of the petiole as a taxonomic character," *Proc. IX Int. Bot. Congress*, **2:** 171.

Johnson, M. A., and Truscott, F. H., 1956. "On the anatomy of *Serjania*. I. The path of the bundles," *Amer. Jour. Bot.*, **43:** 509–518.

Lee, E., 1912. "Observations on the seedling anatomy of certain Sympetalae. I. Tubiflorae," *Ann. Bot.*, **26:** 727–746.

———, 1914. "Observations on the seedling anatomy of certain Sympetalae. II. Compositae," *Ann. Bot.*, **28:** 303–329.

Lenoir, M., 1920. "Evolution du tissu vasculaire chez quelques plantules de dicotylédones," *Ann. Sci. Nat. Bot.*, sér. 10, **2:** 1–123.

Lignier, O., 1887. "Recherches sur l'anatomie comparée des Calycanthées, des Mélastomacées et des Myrtacées," *Arch. Bot. du Nord de la France*, **3:** 1–455.

LIGNIER, O., 1890. "Recherches sur l'anatomie des organes végétatifs des Lecythidacées," *Bull. Sci. Fr. et Belg.*, **21**.

LUBBOCK, J., 1892. *Contribution to our knowledge of seedlings*, New York: Appleton.

MAJUMDAR, G. P., 1941. "Anomalous structure of the stem of *Nyctanthes arbortristis* L.," *Jour. Indian Bot. Soc.*, **20**: 119–122.

MARSDEN, MARGERY P. F., and BAILEY, I. W., 1955. "A fourth type of nodal anatomy in dicotyledons, illustrated by *Clerodendron trichotomum* Thunb.," *Jour. Arnold Arb.*, **36**: 1–49.

METCALFE, C. R., and CHALK, L., 1950. *Anatomy of the dicotyledons*, Oxford: Clarendon Press.

MONEY, LILLIAN L., BAILEY, I. W., and SWAMY, B. G. L., 1950. "The morphology and relationships of the Monimiaceae," *Jour. Arnold Arb.*, **31**: 372–404.

NAST, CHARLOTTE G., and BAILEY, I. W., 1946. "Morphology of *Euptelea* and comparison with *Trochodendron*," *Jour. Arnold Arb.*, **27**: 186–192.

POST, D., 1958. "Studies in Gentianaceae. I. Nodal anatomy of *Frasera* and *Swertia perennis*," *Bot. Gaz.*, **120**: 1–14.

SINNOTT, E. W., and BAILEY, I. W., 1914a. "Investigations on the phylogeny of the angiosperms. I. The anatomy of the node as an aid in the classification of the angiosperms," *Amer. Jour. Bot.*, **1**: 303–322.

——, ——, 1914b. *Ibid.*, III. "Nodal anatomy and the morphology of stipules," *Amer. Jour. Bot.*, **1**: 441–453.

SOLEREDER, H., 1908. *Systematic anatomy of the dicotyledons* (trans. by Boodle and Fritsch), Oxford: Clarendon Press.

SWAMY, B. G. L., 1953. "The morphology and relationships of the Chloranthaceae," *Jour. Arnold Arb.*, **34**: 375–408.

THOMAS, E. N., 1914. "Seedling anatomy of Ranales, Rhoedales, and Rosales," *Ann. Bot.*, **28**: 695–733.

TRONCHET, A., 1930. "Recherches sur les types d'organisation le plus répandus de la plantule des dicotylédones. Leurs principales modifications, leurs rapports," *Arch. de Bot.*, **4**: 1–252.

VAN TIEGHEM, P., 1891. "Sur les tubes criblés extralibériens et les vaisseaux extraligneux," *Jour. de Bot.*, **5**: 117–128.

——, and DOULIOT, H., 1886. "Sur la polystélie," *Ann. Sci. Nat. Bot.*, sér. 7, **3**: 275–322.

WATARI, S., 1934. "Anatomical studies on some leguminous leaves with special reference to the vascular system in petioles and rachises," *Jour. Fac. Sci. Tokyo Univ.*, sect. III, **4**: 225–365.

——, 1936. "Anatomical studies on the vascular system in the petioles of some species of *Acer*, with notes on the external morphological features," *Jour. Fac. Sci. Tokyo Univ.*, sect. III, **5**: 1–73.

WILSON, C. L., 1924. "Medullary bundles in relation to primary vascular system in the Chenopodiaceae and Amaranthaceae," *Bot. Gaz.*, **78**: 175–199.

WORSDELL, W. C., 1919. "The origin and meaning of medullary (intraxylary) phloem in the stems of dicotyledons. II. Compositae," *Ann. Bot.*, **33**: 421–458.

chapter seven ▶▶▶▶▶ Root

Relatively little attention has been paid to anatomy of roots in relation to the taxonomic system. This situation seems to have resulted from (1) incomplete knowledge of variability in root anatomy within a species; (2) lack of studies because good comparable material is more difficult to obtain or less frequently collected than for shoots, flowers, etc.; (3) lack of monographs on comparative anatomy of roots as compared with monographs on anatomy of other organs; and (4) the impression, seemingly widespread among both systematists and anatomists, that root anatomy throughout angiosperms is stereotyped. This chapter will attempt to show that roots are not as stereotyped anatomically as one might assume; nevertheless, attention must be called to anatomical variability among roots of a species.

That root anatomy may be of importance is indicated by the fact that Metcalfe and Chalk (1950) have included this topic under each family for which such information is available. Interestingly, considerable attention is devoted to root anatomy in Solereder and Meyer's *Systematische Anatomie der Monokotyledonen,* and we can expect similar emphasis in Metcalfe's forthcoming volumes on anatomy of monocots. Likewise, a paper by Lindinger (1906) and the chapter on roots in Arber (1925) suggest the variety of anatomical conditions in monocots. The most significant single paper on comparative anatomy of angiosperm roots, however, is that by Guttenberg (1940). A recent bibliography has been provided by Miller (1960).

EPIDERMIS AND CORTEX

Root hairs and epidermis

Linsbauer (1930) has summarized literature on root hairs with respect to a number of anatomical features. Studies by Olivier (1881) and Kroemer (1903) cannot be overlooked in this respect. Van Tieghem (1887) has monographed the curious phenomenon of paired root hairs.

Velamen

Velamen is a multiple epidermis specialized for absorption and storage of water. Families in which it occurs are listed by Guttenberg (1940, 1943) and Arber (1925). Occurrence of "passage cells" as opposed to cells with thickenings may be a systematic criterion (Jancewski, 1885); other criteria include number of layers, relative prominence or wideness of bands, and the patterns formed by these thickenings.

Exodermis

Differentiation of one or more subepidermal layers, usually by means of sclerification or suberization, is characteristic of many angiosperm roots, particularly those of monocots. The best compilations of information on these topics are offered by Guttenberg (1943), Kroemer (1903), and Olivier (1881). Two distinctive exodermis patterns are illustrated in Fig. 7-1A-B.

Cortex

Patterns of cortical parenchyma cells are by no means uniform in angiosperms. Distinct radial arrangement may or may not occur, and parenchyma may take the form of radial or transverse plates. Sclerification of cortex and formation of tanniniferous idioblasts are two modifications shown for two species of Rapateaceae in Fig. 7-1C-D. Tetley (1925) has shown that secretory canals in roots of Compositae offer comparative criteria such as relative size, number, type of development, and location within the cortex. Secretory canals, ethereal oil cells, and laticifers in roots are discussed by Guttenberg (1940). Differentiation of the layer outside the endoder-

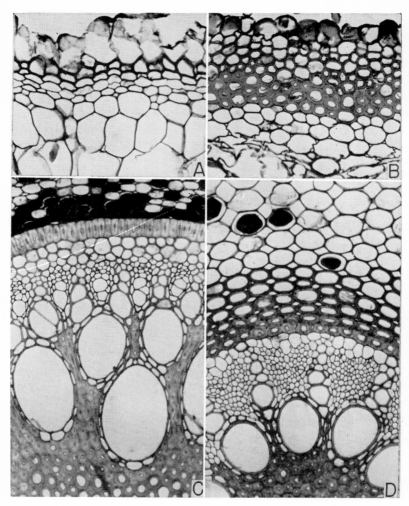

Fig. 7-1. Comparison of outer region (A, B) and inner regions (C, D) of roots for species of Rapateaceae. A, *Duckea flava*, showing layers of smaller, rather thin-walled cells interior to root epidermis; B, *Saxofridericia spongiosa*, showing markedly sclerified cells several layers interior to the epidermis; C, *Guacamaya superba*, showing markedly thick-walled endodermal and pith cells, large vessels extending well into the center of the root; D, *Rapatea Spruceana*, showing a thick-walled endodermis, with cortical cells exterior to it with similar thickenings. Black cells are tanniniferous. All ×140.

mis by means of peculiar thickenings characteristic of particular species is a curious feature of Cruciferae illustrated in Fig. 7-2, based on work of Van Tieghem and Douliot (1888). Idioblasts may occur in roots, and Bloch's (1946) paper on trichoscleids in *Monstera* deserves mention. Weisse (1897) discusses occurrence and structure of lenticels in aerial roots of both monocots and dicots.

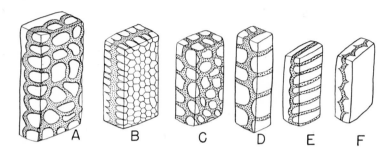

Fig. 7-2. Cells of the layer just outside the endodermis, from roots of various species of Cruciferae, to show various patterns of wall thickenings characteristic of particular taxa. Cells are viewed obliquely; internal face, right, and a radial wall at left in each case. A, *Cheiranthus Cheiri*; B, *Sinapis alba*; C, *Alyssum montanum*; D, *Schizopetalum Walkeri*; E, *Lepidium sativum*; F, *Iberis affinis*. All × *ca.* 400. (Redrawn from Van Tieghem and Douliot, 1888.)

ENDODERMIS

The endodermis, a nearly universal feature of angiosperm roots, may occur in a variety of ways that often show close correlation with the taxonomic system. Guttenberg (1943) recognizes three main types of endodermis: primary, with only a casparian strip; secondary, in which suberization of the cell is general; and tertiary, in which sclerification occurs. Varied expression or combination of these is useful systematically. For example, the distinctive patterns of sclerification in Xyridaceae, shown in Fig. 7-3, prove excellent criteria for species in the genus *Abolboda* (Carlquist, 1960).

VASCULAR CORE

The vascular core offers many features. Number of xylem and phloem poles has been used comparatively, but this varies so widely

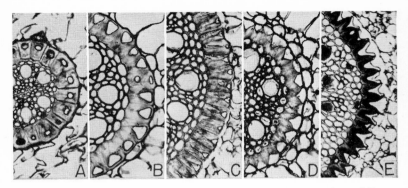

Fig. 7-3. Portions of stelar cylinder of roots of Xyridaceae, to show different patterns of endodermal-cell thickenings characteristic of species and genera. A, *Abolboda acaulis;* B, *A. Sprucei;* C, *A. macrostachya;* D, *A. linearifolia;* E, *Achlyphila disticha.* All ×200. (From Carlquist, 1960.)

within a single plant that its usefulness is doubtful. If markedly different (for example, *Abolboda macrostachya;* see Carlquist, 1960), it may be significant. In monocots, which have a cylinder of xylem poles, additional isolated vessels may be present in the central portion of the vascular core or not (Fig. 7-3), and this feature has been used systematically. Irregular grouping of xylem and phloem, as in *Pandanus* (Guttenberg, 1940), or polystely (White, 1907) are of interest. Phloem may be scattered throughout the vascular core, as in some Xyridaceae (Fig. 7-3) as well as in discrete poles. Important in monocots is the occurrence of so-called mechanical roots, in which sclerenchyma dominates in the vascular core. Solereder (1908) gives a listing of dicot families in which vascular anomalies occur in roots.

Cheadle (1944) has shown that occurrence of vessels and details of their structure are important in monocot roots. The phyletic significance of this is described in Chapter 4. Presence of phloem fibers in roots has been regarded by Van Tieghem (1888) as characteristic of Malvaceae.

LATERAL ROOTS

Mode of formation and histological features of lateral roots would seem unpromising for comparative study. However, two papers (Lemaire, 1886; Van Tieghem and Douliot, 1888) suggest a number of comparative criteria that may be yielded by such investigations.

TYPES OF ROOTS

Roots show anatomical specializations with regard to mesophytic, xerophytic, and hydrophytic habitats, although anatomical structure of roots within one of these categories might seem primarily a matter of ecological anatomy. Comparison of roots of mangrove-type plants is given by Liebau (1913). Contractile roots have attracted much attention, as the studies of Berckemeyer (1929), Nordhausen (1913), Rohde (1928) and Stroever (1892) indicate. Root dimorphism, as in the "feeding" and "climbing" roots of Araceae, deserves further investigation. An instance of root polymorphism is furnished by Majumdar (1932).

HAUSTORIA

Haustoria of parasitic angiosperms, and the nature of their connection to roots of host plants, offer topics of interest to comparative anatomists. An excellent summary on these topics has been provided by Sperlich (1925).

COMPARATIVE STUDIES

A series of studies attests to the systematic usefulness of root anatomy. That of Maxwell (1893) emphasizes differences in patterns of secondary growth. Losch (1913) discusses such features as distribution of laticifers. Kean (1927) uses the following features in the genus *Mesembryanthemum:* presence or absence of anomalous secondary growth; presence or absence of pith; cork formation followed or preceded by anomalous cambium formation. Investigating two species of *Silene,* Millner (1934) found minor but constant differences. Soper (1959) lists a number of features that seem systematic criteria in roots of legumes. Perhaps the most interesting and exhaustive use of anatomical characteristics of roots is that of Woodson (1957) in the genus *Rauwolfia.* By means of such features as phloem sclerenchyma, vessel diameter, and distribution of starch, he is able to construct a key to species based on root anatomy.

Monocot roots, as Arber (1925) suggests in the case of the "piliferous layer," may have a greater variety of anatomical expressions than dicot roots. If true, this might be related to the relative persistence of primary tissues in monocot roots—which rarely have any secondary growth—as compared with those of dicots. Nevertheless, the studies cited in the preceding paragraph all deal with dicots. Root anatomy may, it is true, be more useful at generic and specific levels than in broader comparisons, but comparative anatomical studies of this organ should be encouraged.

REFERENCES

ARBER, AGNES, 1925. *Monocotyledons,* Cambridge: Cambridge University Press.

BERCKEMEYER, W., 1929. Über kontraktile Umbelliferenwurzeln," *Bot. Arch.,* 24: 273–318.

BLOCH, R., 1946. "Differentiation and pattern in *Monstera deliciosa.* The idioblastic development of the trichosclereids in the air root," *Amer. Jour. Bot.,* 33: 544–551.

CARLQUIST, S., 1960. "Anatomy of Guayana Xyridaceae: *Abolboda, Orectanthe,* and *Achlyphila,*" *Mem. N. Y. Bot. Gard.,* 10: 65–117.

CHEADLE, V., 1944. "Specialization of vessels within the xylem of each organ in the Monocotyledoneae," *Amer. Jour. Bot.,* 31: 81–92.

GUTTENBERG, H. VON, 1940. "Der primäre Bau der Angiospermenwurzel," in K. Linsbauer, ed., *Handbuch der Pflanzenanatomie* 8(3): viii + 369.

———, 1943. "Die physiologischen Scheiden," in K. Linsbauer, ed., *Handbuch der Pflanzenanatomie* 5(4): viii + 217.

JANCZEWSKI, E. DE, 1885. "Organisation dorsiventrale dans les racines des Orchidées," *Ann. Sci. Nat. Bot.,* sér. 7, 2: 55–81.

KEAN, C. I., 1927. "Anatomy of the genus *Mesembryanthemum.* I. Root," *Trans. Bot. Soc. Edinburgh,* 29: 381–388.

KROEMER, K., 1903. "Wurzelhaut, Hypodermis, und Endodermis der Angiospermenwurzel," *Bibl. Bot.,* 59: 1–151.

LEMAIRE, A., 1886. "Recherches sur l'origine et le développement des racines latérales chez les dicotylédones," *Ann. Sci. Nat. Bot.,* sér. 7, 3: 163–274.

LIEBAU, O., 1913. "Beiträge zur Anatomie und Morphologie der Mangrove-Pflanzen, insbesondere ihres Wurzelsystems," *Beitr. Biol. Pfl.,* 12: 182–213.

LINDINGER, L., 1906. "Zur Anatomie und Biologie der Monokotylenwurzel," *Beih. Bot. Centr.,* 19(1): 321–358.

LINSBAUER, K., 1930. "Die Epidermis," in K. Linsbauer, ed., *Handbuch der Pflanzenanatomie,* 4: vii + 277.

Losch, H., 1913. "Beiträge zur vergleichenden Anatomie der Urticineen-wurzeln mit Rücksicht auf die Systematik," thesis, Göttingen.

Majumdar, G. P., 1932. "Heteroarchic roots in *Enhydra fluitans* Lour.," *Jour. Indian Bot. Soc.*, **11**: 225–227.

Maxwell, F. B., 1893. "A comparative study of the roots of the Ranunculaceae," *Bot. Gaz.*, **18**: 8–16, 41–47.

Miller, R. H., 1960. *Morphology and anatomy of roots*, New York: Scholar's Library.

Millner, M. E., 1934. "Anatomy of *Silene vulgaris* and *Silene maritima* as related to ecological and genetical problems. I. Root structure," *New Phyt.*, **33**: 77–95.

Nordhausen, M., 1913. "Über kontraktile Luftwurzeln," *Flora*, **105**: 101–126.

Olivier, L., 1881. "Recherches sur l'appareil tegumentaire des racines," *Ann. Sci. Nat. Bot.*, sér. 6, **11**: 5–133.

Rohde, H., 1928. "Über die kontraktilen Wurzeln einiger Oxalidaceen," *Bot. Arch.*, **22**: 463–532.

Solereder, H., 1908. *Systematic anatomy of the dicotyledons* (trans. by Boodle and Fritsch), Oxford: Clarendon Press.

Soper, Kathleen, 1959. "Root anatomy of grasses and clovers," *N. Z. Jour. Agr. Res.*, **2**: 329–341.

Sperlich, A., 1925. "Organe besonderer physiologischer Dignität," in K. Lins-bauer, ed., *Handbuch der Pflanzenanatomie*, **9**(2): iv + 52.

Stroever, V., 1892. "Über die Verbreitung der Wurzelkürzung," thesis, Jena.

Tetley, Ursula, 1925. "The secretory system of roots of the Compositae," *New Phyt.*, **24**: 138–162.

Van Tieghem, P., 1887. "Sur les poils radicaux geminés," *Ann. Sci. Nat. Bot.*, sér. 7, **6**: 127–128.

————, 1888. "Sur les fibres libériennes primaires de la racine des Malvacées," *Ann. Sci. Nat. Bot.*, sér. 7, **7**: 176.

————, and Douliot, H., 1888. "Recherches comparatives sur l'origine des membres endogènes dans les plantes vasculaires," *Ann. Sci. Nat. Bot.*, sér. 7, **8**: 1–660.

Weisse, A., 1897. "Ueber lenticellen und verwandten Durchlüftungseinrich-tungen bei Monokotylen," *Ber. Deutsch. Bot. Ges.*, **15**: 303–320.

White, J. H., 1907. "On polystely in roots of Orchidaceae," *Univ. Toronto Stud.*, Biol. ser., **6**: 1–20.

Woodson, R. E., Jr., 1957. "The botany of *Rauwolfia*," in *Rauwolfia*, by R. E. Woodson, H. W. Youngken, E. Schlitter, and J. A. Schneider. Boston: Little, Brown.

chapter eight

▶▶▶▶▶ Histology
of the Stem

PRIMARY STEM

The primary stem is an arbitrarily defined portion of the
shoot system, a mere *coupe caractéristique* in the ontogeny of the
shoot. This is mentioned because not all tissues of the primary stem
are correlated in their maturation with maturation of primary
xylem, which defines the primary stem. For example, sclerenchyma
of stems typically matures after secondary growth has begun, but
may conveniently be considered under this heading. Although study
of the vascular pattern of the stem is a three-dimensional matter,
studies of other aspects of the stem have usually been made by means
of arbitrary—or random—sections (usually transections). There is no
danger in use of such sections unless material that is not truly
comparable by virtue of maturity or position in the plant is offered
as comparable.

Dicotyledons

Vascular bundles. Among dicots, primary vascular tissue as
seen in transection may be composed of a nearly continuous cylinder
or of discrete bundles. This difference, long recognized, does not
correlate well with differences in growth habit or phylogenetic
advancement, for exceptions in these respects are rather numerous.
The controversy concerning such interpretations of the angiosperm
stele, involving such matters as Jeffrey's "foliar ray hypothesis," has

attracted the attention of Bailey, Sinnott, Thoday, Petersen, and others. Opinions of these workers are ably summarized by Metcalfe and Chalk (1950). Although many elements of this controversy can now be forgotten, the work of Barghoorn, cited in Chapter 4, has clarified the nature of the primary vascular cylinder and its altera- tions during secondary growth. Metcalfe and Chalk (1950) list families in which pith rays are broad, narrow, or mixed. Presence of a continuous cylinder versus separate bundles, although no longer of major phylogenetic significance, may well be of taxonomic interest in some groups. Kostyschew (1924) gives much comparative data on these points, especially in relation to stem sclerenchyma.

A matter that is related, in that it involves capability of bundles for secondary growth, is the question of whether the primitive habit in angiosperms is herbaceous or woody. Sinnott and Bailey (1914) hypothesized that modern herbs are advanced over woody forms. With very few exceptions, herbs show more advanced xylem struc- ture than the woody members of their respective families or orders. This is true despite the fact that primary xylem—which is present in all herbs although secondary xylem may not be—tends to be a "refugium" of primitive expressions and is conservative compared with secondary xylem in dicots (see Chapter 4). Cheadle (1953) gives a series of compelling arguments that suggest that ancestors of both monocots and dicots would be classified as woody dicots. There is still the possibility, which Cheadle admits, that a herbaceous "pre- monocot" could have given rise to monocots and subsequently vanished. Most important about the data concerning these hypoth- eses, however, is the inescapable conclusion that within dicots, the herb is an advanced form. By inference, we may more logically conclude that dicot ancestors were woody than that they were herbs. However, the writer believes that the trend to herbaceousness is not completely irreversible. It seems unlikely that no herbs have been able to increase production of secondary xylem.

Collenchyma. In both stems and petioles, collenchyma is a tissue that may be of systematic importance. In addition to presence or absence, the number of cell layers, occurrence as ridges or as a continuous sheath, hypodermal or deep-seated location, and nature of cell contents were listed by Vesque (1875) as features worth consideration. The pattern of collenchymatic thickenings as seen in transection was emphasized by Müller. Little change in his terminology occurred prior to Duchaigne's (1952) demonstration that centrifugal phenomena should also be classified as collenchyma. In these phenomena, pectic materials—presumably similar to those incorporated centripetally in walls of "typical" collenchyma—are extruded into intercellular spaces and have much the same appearance and perhaps function as "classical" centripetal collenchyma. Examples of centrifugal collenchyma have been found in *Aucuba japonica* and *Nerium oleander* (Duchaigne, 1952) as well as in

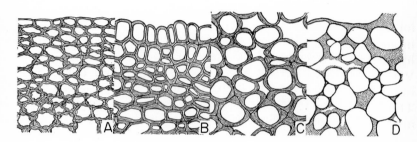

Fig. 8-1. Types of collenchyma in stems of dicotyledons. A, angular collenchyma (thickenings in corners—*Cucurbita pepo;* B, tangential collenchyma (thickenings greatest on radial walls)—*Sambucus velutina;* C, lacunar collenchyma (collenchyma containing large intercellular spaces)—*Fitchia speciosa;* D, centrifugal collenchyma (pectic materials extruded into intercellular spaces)—*Madia sativa.* All × *ca.* 200.

Madia sativa (Carlquist, 1959b). Examples of the three types of centripetal collenchyma, as well as the centrifugal sort, are shown in Fig. 8-1. The possibility of intermediacy between centrifugal and centripetal types is suggested by the ontogeny of collenchyma in *Heracleum* (Majumdar and Preston, 1941) in which centrifugal deposition fills "potential intercellular spaces" early in collenchyma development, although most thickening of walls in mature collenchyma cells is a result of centripetal deposition. Regardless of what types are recognized, the most important features of collenchyma to the systematist may be the distribution and abundance in the plant body.

Parenchyma. Vesque (1875) noted that such items as number of cell layers, presence of a hypodermis, relative amounts of photosynthetic and nonphotosynthetic tissue, and development of a starch sheath can prove useful comparative features in cortical parenchyma.

Transformation of ground tissue cells of cortex into transfusion cells forms an interesting taxonomic feature in some groups such as *Casuarina* (Metcalfe and Chalk, 1950).

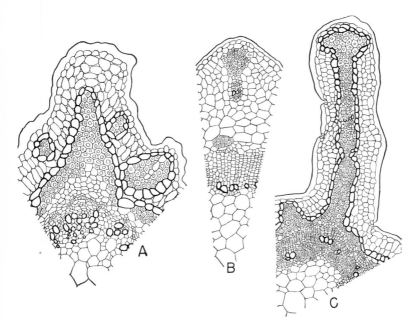

Fig. 8-2. Comparison of stem structure in three species of *Genista*. Note differences in sclerenchyma distribution and shape of fiber areas. A, *Genista ulicina;* B, *G. tinctoria;* C, *G. pteroclada.* (Redrawn from Pellegrin, 1908.)

Sclerenchyma. Blyth (1958) has shown in selected cases that distinctions between phloem fibers, protophloem fibers, bundle cap fibers, and cortical fibers are not easy to draw in many instances, although extreme conditions may be easily interpreted. Comparative data on fibers have been provided by Morot (1885) and Tobler (1939, 1957), and many monographs include such data. Particularly noteworthy are the studies of Nestler (1932) on *Linum* and Pellegrin (1908) on *Genista* and related genera. From the latter study, three figures have been redrawn (Fig. 8-2) to show distinctive aspects in fiber distribution. Cortical sclerification can proceed so far that inner cortex, pith rays, and even cambial region consist of sclerenchyma, as shown by *Goodenia ovata* (Vesque, 1875), *Calycadenia tenella* (Carlquist, 1959b), and other taxa (Kostyschew, 1924). These conditions can be systematically useful, as can the lack of sclerenchyma in these regions, for which condition a listing of families is given by Metcalfe and Chalk (1950). Dicot families with cortical fibers and "pericyclic" fibers are listed by Solereder (1908).

Endodermis. Stem endodermis is present in various ways in a variety of dicots. It may take the form of a single layer of cells with casparian strips, a layer of sclereids, or merely a layer of cells of different size, or lacking contents, or with weak radial walls. For accounts of stem endodermis, which may be a systematic criterion, the reader is referred to Guttenberg (1943) and the accounts of Metcalfe and Chalk (1950) under Piperaceae, Haloragaceae, Labiatae, and Compositae.

Pith. Gris (1872) has summarized systematically useful features of pith:

Parenchyma: alive (indicated by presence of starch) or dead; parenchyma shrunken, forming a pith cavity or cavities; size of cells; shape, especially as related to intercellular spaces (round, polygonal, spongy); presence of crystals, mucilage, etc.; parenchyma diaphragms.

Sclerenchyma: wall thickness; nests of sclereids; reticulate pattern of sclereids; sclerenchyma diaphragms; sclerenchyma adjacent to inner faces of vascular bundles.

For listing of families with pith diaphragms, see Solereder (1908) or Metcalfe and Chalk (1950). A study by Flot (1893) of pith areas immediately adjacent to inner faces of vascular bundles reveals features of taxonomic importance. Solereder (1908) mentions dicot families in which pith sclerification occurs.

A systematically significant array of pith types occurs within the genus *Dubautia* (Carlquist, 1959a). Likewise, comparison of the species of *Fitchia* (Fig. 8-3) reveals anatomical differences by means of which each species may be identified.

Special stem types. Although Metcalfe and Chalk (1950) summarize a large quantity of material on stem anatomy, monographs of particular types may be helpful. Among such studies are Boosfeld's (1920) on stem succulents and Costatin's (1884) on aquatic stems.

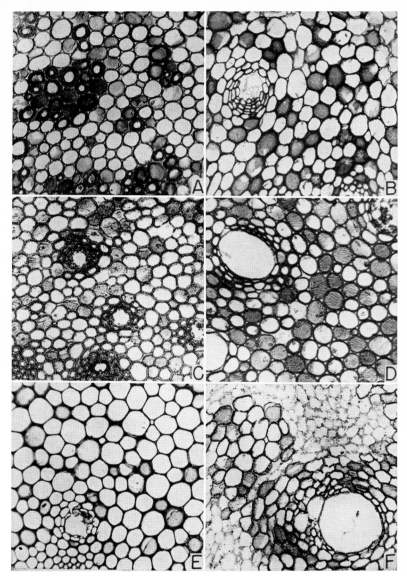

Fig. 8-3. Types of pith structure that characterize species of *Fitchia* (Compositae). Note presence or absence of secretory canals, presence of thin-walled and thick-walled cells, shape of cells, nature of pitting, and types of cells surrounding secretory canals. A, *F. cuneata;* B, *F. cordata;* C, *F. tahitensis;* D, *F. nutans;* E, *F. speciosa;* F, *rapensis.* All × *ca.* 140.

107

Monocotyledons

Although never completed, Solereder and Meyer's *Systematische Anatomie der Monokotyledonen* contributed much to our knowledge of monocot stems, as have the monographs of Wille (1916) and Arber (1925, 1934). Much additional information will soon be available in the volumes of Metcalfe's *Anatomy of the Monocotyledons*.

Cheadle has developed a particularly valuable tool in his studies of tracheary elements in monocot stems (see Chapter 4). For sum-

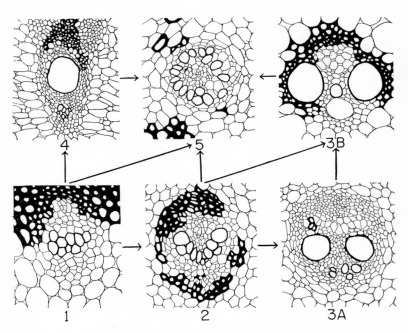

Fig. 8-4. Types of bundle structure in monocots and their phylogenetic interrelationships. Type 1, simple collateral bundle (*Liriope Muscari*); Type 2, collateral bundle with a V-shaped xylem (*Dracaena fragrans*); Type 3A, collateral bundle with two large metaxylem elements (*Acoelorrhaphe Wrightii*); Type 3B, "graminean type," collateral bundle in which xylem and phloem meet along a curved or nearly V-shaped line (*Arundo donax*); Type 4, collateral bundle with one large metaxylem element and phloem that forms an inverted U-shape (*Roystonea regia*); Type 5, amphivasal bundle (*Cordyline terminalis*). Sclerenchyma shown in black, xylem elements shown in bold outline. (Drawn from Cheadle and Uhl, 1948.)

mary of distribution of tracheid and vessel types in monocots, the reader is referred to his 1944 paper and the papers cited therein. Cheadle and Uhl (1948) ingeniously used phylogenetic data from tracheary elements to develop ideas concerning phylogeny of bundle types in monocot stems and leaves. The six types and their phylogenetic interrelations are shown in Fig. 8-4.

Taxonomic usefulness of the vascular core of monocot stems is admirably illustrated by Ogden (1943) in the genus *Potamogeton*. Although he does not designate a phylogenetic relationship among the types (certainly the single-bundle type must be highly advanced), the taxonomic utility of stem types for discriminating among species in this genus is obviously quite considerable.

PERIDERM AND BARK

The work of Douliot (1889) on periderm is significant because it gives data, arranged systematically, on the site of origin of periderm for a large number of genera. Metcalfe and Chalk (1950) have questioned whether superficial versus deep-seated origin characterize natural groups; both types may occur in a single family. Nevertheless, this feature has some value, as the listings of Solereder and of Metcalfe and Chalk suggest. Zahur (1959) gives comparative data on periderm of woody dicots. He considers such features as nature of phelloderm, extent of periderm, and presence of layering. Douliot (1889) gives examples of exceptional suberization or lignification in periderm.

For a comparative study of lenticels, the reader may consult Devaux (1900). Wetmore (1926) has found phylogenetic significance in lenticels, and he described evolutionary alteration from transversely to vertically elongate structures.

Anatomy of bark is summarized under accounts of dicot families by Metcalfe and Chalk (1950). Worthy of special mention is the work of Chattaway (1953) on *Eucalyptus* bark. In addition to presenting descriptive material, she expresses in chart form such features as persistence of bark, wideness of phellem, presence of expanded parenchyma and parenchyma wedges, and radially elongate phelloderm cells. In a series of papers, she monographs such aspects as oil glands (1955a), enlarged fibers (1955b), and radially elongate phelloderm cells (1955c). These studies are models for clarity of approach, presentation by systematic groupings, and interest in developmental aspects. Zahur's (1959) monograph indicates that presence or absence of sclerenchyma, time of origin of bark (near cambium or in inactive phloem), nature of sclerenchyma walls (sculptured, gelatinous), presence of septate fibers, and association of sclerenchyma with other cell types are among the gamut of expressions worthy of attention for their systematic possibilities.

ANOMALOUS SECONDARY THICKENING

This feature occurs in many families of dicots. For listing of these, see Boureau (1957), Metcalfe and Chalk (1950), and Pfeiffer (1926). Logically, Pfeiffer focuses attention on the nature of the anomaly, systematic occurrence, relation to habit of the plant, and developmental history. Often, anomalous secondary thickening proves related to particular growth forms, such as lianas, although this is by no means always true. The study of Bailey and Howard (1941) shows the usefulness of anomalous tissue distributions as taxonomic criteria when combined with other characteristics. There can be little doubt that means of anomalous secondary thickening are derived from normal conditions. The transition to abnormal thickening appears to have taken place in a number of groups of dicots, and both normal and abnormal modes may occur in a single genus. Monographs dealing with related phenomena include those of Dusén and Neger (1921) on basal thickenings of a caudexlike nature and Moss and Gorham (1953) on fission of stems into irregular segments by means of interxylary cork formation.

SECONDARY GROWTH IN MONOCOTYLEDONS

Information on production of bundles from a thickening ring or "monocot cambium" is ably summarized by Pfeiffer (1926), who discusses Liliaceae, Amaryllidaceae, and several other families. Cheadle (1937) has added comparative information on bundle type, growth rings, and means of vascular connection between leaf traces and secondary bundles. Pfeiffer (1926) offers a phylogenetic inter-pretation—namely, the possibility that occurrence of secondary bundles in monocots may be related to supposed antiquity of the groups that possess them. In addition to formation of secondary bundles by means of a thickening ring, cambial activity can also occur within monocot bundles (Arber, 1919).

INFLORESCENCE AXIS

Anatomy of inflorescence axes has not been an object of much comparative study, although Solereder and Meyer's *Systematische Anatomie der Monokotyledonen* suggests that considerable use could be made of such features as a sclerenchyma cylinder, its thickness, arrangement of bundles and their type, etc. Cheadle (1944), and literature he cites, has shown the phylogenetic significance of vessel occurrence and nature in inflorescence axes. Arber (1934, Chap. 9) has shown the value of inflorescence-axis anatomy in systematic study of Gramineae. On the whole, however, little has been done, in many groups, to continue investigation of the types of anatomical data studied by Reiche (1887).

REFERENCES

ARBER, AGNES, 1919. "Studies in intrafascicular cambium in monocotyledons," *Ann. Bot.*, **33**: 459–465.

———, 1925. *Monocotyledons*, Cambridge: Cambridge University Press.

———, 1934. *The Gramineae*, Cambridge: Cambridge University Press.

BAILEY, I. W., and HOWARD, R. A., 1941. "The comparative morphology of the Icacinaceae. I. Anatomy of the node and internode," *Jour. Arnold Arb.*, **22**: 125–132.

BLYTH, AMELIE, 1958. "Origin of primary extraxylary fibers in dicotyledons," *Univ. Calif. Publ. Bot.*, **30**: 145–232.

BOOSFELD, A., 1920. "Beiträge zur vergleichenden Anatomie stammsukkulenter Pflanzen," *Beih. Bot. Centr.*, **37**(1): 217–258.

BOUREAU, E., 1957. *Anatomie végétale*, Vol. 3, Paris: Presses Universitaires de France.

CARLQUIST, S., 1959a. "Vegetative anatomy of *Dubautia, Argyroxiphium,* and *Wilkesia* (Compositae)," *Pac. Sci.*, **13**: 195–210.

———, 1959b. "Studies on Madinae: anatomy, cytology, and evolutionary relationships," *Aliso*, **4**: 171–236.

CHATTAWAY, M. MARGARET, 1953. "The anatomy of bark. I. The genus *Eucalyptus*," *Austral. Jour. Bot.*, **1**: 402–433.

———, 1955a. *Ibid.*, II, "Oil glands in *Eucalyptus* species," *Austral. Jour. Bot.*, **3**: 21–27.

Chattaway, M. Margaret, 1955b. *Ibid.*, III. "Enlarged fibers in the bloodwoods *(Eucalyptus* spp.) ," *Austral. Jour. Bot.,* **3**: 28–38.

———, 1955c. *Ibid.*, IV. "Radially elongated cells in the phelloderm of species of *Eucalyptus,*" *Austral. Jour. Bot.,* **3**: 39–47.

Cheadle, V., 1937. "Secondary growth by means of a thickening ring in certain monocotyledons," *Bot. Gaz.,* **98**: 535–555.

———, 1944. "Specialization of vessels within the xylem of each organ in the Monocotyledoneae," *Amer. Jour. Bot.,* **31**: 81–92.

———, 1953. "Independent origin of vessels in the monocotyledons and dicotyledons," *Phytomorphology,* **3**: 23–44.

———, and Uhl, Natalie W., 1948. "Types of vascular bundles in the Monocotyldoneae and their relation to the late metaxylem conducting elements," *Amer. Jour. Bot.,* **35**: 486–496.

Costatin, J., 1884. "Recherches sur la structure de la tige des plantes aquatiques," *Ann. Sci. Nat. Bot.,* sér. 6, **19**: 287–331.

Devaux, H., 1900. "Recherches sur les lenticelles," *Ann. Sci. Nat. Bot.,* sér. 8, **12**: 1–240.

Douliot, H., 1889. "Recherches sur la périderme," *Ann. Sci. Nat. Bot.,* sér. 7, **10**: 325–395.

Duchaigne, A., 1952. "L'ontogénie des collenchymes chez les dicotylédones," *C. R. Acad. Sci.,* **234**: 1903–1905.

Dusén, P., and Neger, F. W., 1921. "Über Xylopodien," *Beih. Bot. Centr.,* **38**(2): 258–317.

Flot, L., 1893. "Recherches sur la zone périmédullaire de la tige," *Ann. Sci. Nat. Bot.,* sér. 7, **18**: 37–112.

Gris, A., 1872. "Sur la moelle des plantes ligneuses," *Ann. Sci. Nat. Bot.,* sér. 5, **14**: 34–79.

Guttenberg, H. von, 1943. "Die physiologischen Scheiden," in K. Linsbauer, ed., *Handbuch der Pflanzenanatomie,* **5**(4): viii + 217.

Kostyschew, S., 1924. "Der Bau und das Dickenwachstum der Dikotylenstämme," *Beih. Bot. Centr.,* **40**(1): 295–373.

Majumdar, G. P., and Preston, R. D., 1941. "The fine structure of collenchyma cells in *Heracleum sphondylium* L.," *Proc. Roy. Soc. Lond.,* ser. B, **130**: 201–217.

Metcalfe, C. R., and Chalk, L., 1950. *Anatomy of the dicotyledons,* Oxford: Clarendon Press.

Morot, L., 1885. "Recherches sur le péricycle . . . ," *Ann. Sci. Nat. Bot.,* sér. 6, **20**: 217–309.

Moss, E. H., and Gorham, Anne L., 1953. "Interxylary cork and fission of stems and roots," *Phytomorphology,* **3**: 285–294.

Nestler, H., 1932. "Beiträge zur systematischen Kenntnis der Gattung *Linum,*" *Beih. Bot. Centr.,* **50**(2): 497–551.

Ogden, E. C., 1943. "The broad-leaved species of *Potamogeton* of North America north of Mexico," *Contrib. Gray Herb.,* **147**: 57–216.

Pellegrin, F., 1908. "Recherches anatomiques sur la classification des genêts et des cytises," *Ann. Sci. Nat. Bot.,* sér. 9, **7**: 129–320.

PFEIFFER, H., 1926. "Das abnorme Dickenwachstum," in K. Linsbauer, ed., *Handbuch der Pflanzenanatomie*, **9:** xiii + 268.

REICHE, K., 1887. "Beiträge zur Anatomie der Inflorescenzaxen," *Ber. Deutsch. Bot. Ges.*, **5:** 310–318.

SINNOTT, E. W., and BAILEY, I. W., 1914. "Investigations on the phylogeny of the angiosperms. IV. The origin and dispersal of herbaceous angiosperms," *Ann. Bot.*, **28:** 547–600.

SOLEREDER, H., 1908. *Systematic anatomy of the dicotyledons* (trans. by Boodle and Fritsch), Oxford: Clarendon Press.

TOBLER, F., 1939. "Die mechanischen Elemente und das mechanische System," in K. Linsbauer, ed., *Handbuch der Pflanzenanatomie*, **4**(6): 1–56.

————, 1957. *Ibid.*, ed. 2, **4**(6): 1–60.

VESQUE, J., 1875. "Mémoire sur l'anatomie comparée de l'écorce," *Ann. Sci. Nat. Bot.*, sér. 6, **2:** 82–198.

WETMORE, R., 1926. "Organization and significance of lenticels in dicotyledons. I. Lenticels in relation to aggregate and compound storage rays in woody stems. Lenticels and roots," *Bot. Gaz.*, **82:** 71–88.

WILLE, F., 1916. "Anatomisch-physiologische Untersuchungen am Gramineen-rhizom," *Beih. Bot. Centr.*, **33:** 1–70.

ZAHUR, M. Z., 1959. "Comparative study of secondary phloem of 423 species of woody dicotyledons belonging to 85 families," *Cornell Univ. Agr. Exp. Stat. Mem.*, **358:** 1–160.

chapter nine ▸ Leaf

The leaf is perhaps anatomically the most varied organ of angiosperms, and its anatomical variations often concur closely with generic and specific—occasionally familial—lines. Therefore, a large quantity of literature has developed. The works cited below represent only a sampling of those in which leaf anatomy predominates. Most often, leaf anatomy is combined with anatomy of other structures in systematic anatomical studies.

LEAF HISTOLOGY

The older works of De Bary, Goebel, and others cover many aspects of leaf organography. In monocots, Arber (1925) combines this approach with histological studies. Systematically valuable histological features and references dealing with comparative anatomy of particular portions of leaves may be listed as follows:

1. *Mesophyll:* bifacial or isolateral (Metcalfe and Chalk, 1950); centric construction with relation to bundles (Metcalfe, 1956; Metcalfe and Chalk, 1950; Stebbins, 1956); mesophyll undifferentiated (homogeneous) into palisade and spongy tissue (Metcalfe and Chalk, 1950); shape of palisade and spongy cells (Meyer, 1923); mesophyll cells with thickenings (Meyer, 1923) or converted into sclereids (Carlquist, 1958; Chartschenko, 1932); secretory canals, cavities, or cells in leaves (Carlquist, 1957a; Chatin, 1875); mesophyll idioblasts present (see Chapter 3).

114

2. *Vein Sheaths:* sheaths present or absent; sheath cells conspicuously large (Metcalfe and Chalk, 1950; Solereder, 1908); cells large and with chloroplasts (Moser, 1934); sclerenchyma present in bundle sheaths (Bailey and Nast, 1944; Carlquist, 1957a, 1958); endodermis present in bundle sheaths (Trapp, 1933; see also Halorageae and Acanthaceae in Metcalfe and Chalk, 1950); presence of cristarque cells (Solereder, 1908).

3. *Hydathodes:* presence, absence, size (Kurt, 1929; Solereder, 1908); hydathode mechanisms not associated with vein endings (Haberlandt, 1894).

4. *Nectaries:* foliar nectaries present or absent, of a distinct form or generalized; with palisade or not, superficial or deeply embedded; (for summary of entire topic, see Zimmermann, 1932).

5. *Stomatal Crypts:* presence or absence (a subgeneric criterion in *Ceanothus;* McMinn, 1942); size, details of construction (Morley, 1953a, 1953b).

6. *Abscission Layers:* time of formation, presence or absence of suberization; location within a leaf; for review, see Pfeiffer, 1928.

VENATION PATTERNS

Venation in three dimensions

Leaves of many species have revolute margins, but this revolution produces no major change in the venation system. In some angiosperms, condensed leaf form is more deeply seated and results from particular types of ontogeny. These types can be of taxonomic and phyletic significance. Leaves of Frankeniaceae (Leinfellner, 1959) show a type of abaxial meristematic development that has also been described for some Ericaceae by Hagerup (1953). Some Pacific Coast Madinae (Carlquist, 1959b) have inverted bundles near the margin owing to a peculiar ontogenetic history. *Argyroxiphium* (Carlquist, 1957b) shows a complicated leaf ontogeny incorporating three series of bundles—a useful systematic criterion. Likewise, the complicated venation of *Agave* leaves (Müller, 1909) must reflect a special type of ontogeny. Examples such as these suggest that comparative studies must often involve developmental approaches, because venation patterns of these sorts may arise in different ways, and interpretation of morphological nature of venation is enhanced by knowledge of ontogenetic history. The occurrence of another peculiar feature of leaf venation, vertically transcurrent veins, is listed for dicot families by Solereder (1908).

Venation of phyllodes provides a special case. Presence of inverted bundles, or of a cylindrical type of bundle organization in phyllodes, represents a specialization of phylogenetic interest. Arber (1925) has utilized such patterns of venation in monocot leaves to suggest a phyllode origin for leaves in this group. Whether leaves of monocots, or some monocots, have originated from phyllodes or not, the collection of these types of data is important both in developing theories of leaf phylogeny and in creating comparisons of value to systematics.

Major venation patterns

Patterns of gross venation, as seen in two dimensions, are of great interest both in monocots and dicots. Unfortunately, little has been done to utilize these in classification, despite the obvious interest of gross venation in such families as Melastomaceae. Kerner and Oliver (1895) have offered a classification that should be considered by any student working with venation patterns. Subtle variations in leaf venation have long been useful for their diagnostic value to paleobotanists, and the early work of Ettinghausen (1861) is still useful to students of leaf venation.

In matters of phylogeny, major venation patterns may prove of increasing significance. Foster (1959) has suggested that the open dichotomous venation of a ranalean genus, *Kingdonia*, may be primitive within angiosperms, and this idea seems worthy of consideration. On the other hand, Takhtajan (1959) claims that leaves with pinnate venation—"open" rather than "closed"—are primitive in angiosperms, and that pinnate-arcuate venation is primitive in monocots.

Minor venation

In leaves of Quiinaceae, both major and minor veins of all genera show a remarkable "lineolate" pattern. Foster, in a series of papers on this family (1950a, 1950b, 1951), has shown that careful description of venation, together with studies of other details of leaf anatomy, can yield valuable taxonomic dividends. The four species of Rubiaceae in the accompanying illustration (Fig. 9-1) suggest that distinctive patterns, when properly analyzed, will be

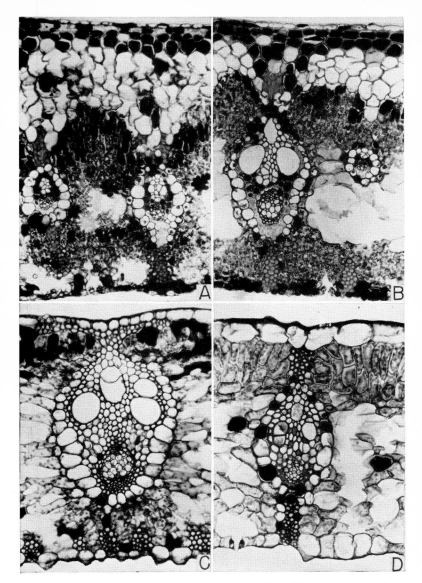

Fig. 9-1. Transections of leaves of Rapateaceae. Note differences among these taxa, including hypodermis presence (above) or absence, types and shapes of chlorenchyma cells, types of bundle structure, and distribution of sclerenchyma. A, *Schoenocephalum coriaceum;* B, *Guacamaya superba;* C, *Rapatea paludosa;* D, *Rapatea Spruceana.* All × *ca.* 35.

117

useful in this group. In certain Hawaiian Compositae (Carlquist, 1959a), minor venation provides taxonomic and evolutionary clues. For information on venation of bud scales, see Müller (1944). Quantitative analysis of minor venation may offer many pitfalls (see Chapter 1). Schuster (1910) has suggested that distance between longitudinal veins and cross veins may be useful. His use of total vein length within a square millimeter might be questioned in view of the work of Zalenski (1902), who shows ecological causes for variation in venation measurements and claims that differing measurements "bear no relation to the taxonomic system." Obviously, use of both major and minor venation is in its infancy. The work of Pray (1955) suggests that distinctive patterns of minor-vein ontogeny, as well as mature venation, can bear comparative analysis.

TYPES OF LEAVES

Leaves may—or may not—show a close relationship with the habitat in anatomical structure. The possibility of evolution along such lines should certainly be entertained in a comparative study. Summarization of "ecological leaf anatomy" is beyond the scope of this book, but certain papers that might fall in this category give useful data. For example, the early monographs of Costatin (1886) and Sauvageau (1891) provide extensive descriptions of leaves of aquatic plants.

A number of monographs on anatomy of special types of leaves have appeared. Among these might be mentioned Godfrin's (1884) work on cotyledon venation and histology, that of Colomb (1887) on stipules, and papers by Leclerc du Sablon (1887), Moens (1956), and Schnee (1939) on tendrils. For information on anatomical structures underlying various types of motion in leaves, see Guttenberg (1926). The palm leaf represents an intriguing special foliar type deserving much more study, and the literature on this topic and its systematic implications are summarized by Eames (1953).

SYSTEMATIC STUDIES

During the last decades of the nineteenth century and the first quarter of the twentieth, numerous monographs on comparative anatomy of leaves—usually organized according to a family—

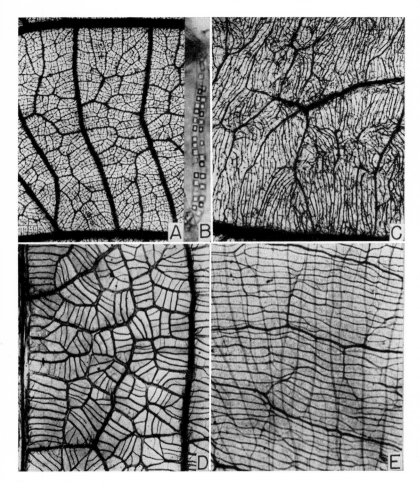

Fig. 9-2. Examples of leaf venation in Rubiaceae, a problem currently under study by Dr. Thomas R. Pray. A, *Guettarda Combsii*, an "ordinary" reticulate pattern; B, same, a crystalliferous trichome from this leaf; C, *Timonius flavescens*, showing a tendency for veins to be oriented at right angles to the leaf axis (left to right); fibers and fibrosclereids form faint streaks in photo; D, *Sommera subcordata*, in which particular areas ("idiodromes") show orientation of veins in certain directions; note rarity of freely terminating veins; E, *Pentagonia spathicalyx*, in which two (largely independent) series of veins, one above the other, run in different directions (producing the "woven" appearance). A, ×6; B, ×320; C–E, ×9.

appeared. Although the vogue for production of these interesting essays (which often incorporate anatomical keys to genera and species) has waned, the value of such studies is as great as it ever has been. Among the many deserving of mention are Priemer's (1893) study on Ulmaceae, Vesque's (1883) on Caryophyllaceae, and Edelhoff's (1886) on Olacaceae. Recent studies on dicots that emphasize taxonomic importance include those of Morley (1953a, 1953b) on *Mouriri* and that of Hagerup (1953) on Ericaceae. In monocots, systematic implications of leaf anatomy have received considerable attention. The work of Müller (1909) and Wunderlich (1950) on *Agave* and the study by Fahn (1954) on Xanthorrhoeaceae are notable. Tomlinson's studies of Zingiberaceae (1956) and Musaceae (1959) use many subtle anatomical features. Smithson (1956) notes similarity in foliar structure between Flagellariaceae and Gramineae. Usefulness of leaf anatomy in Gramineae is suggested by the work of Hubbard (1948) and Stebbins (1956). De Wet (1954) was able, on this basis, to offer conclusions concerning the systematic position of *Danthonia* and its allies. In the accompanying illustration (Fig. 9-2), the complicated structure of leaves in four species of Rapateaceae shows a wealth of features that may be correlated with taxonomic distinctions in this family.

Attention is called to the fact that some of the best examples of systematic and evolutionary use of anatomical features of leaves are described in combination with anatomical studies of other organs, and therefore consultation of the compilations and references of Metcalfe and Chalk (1950) and Solereder (1908), as well as the forthcoming volumes of Metcalfe's *Anatomy of the Monocotyledons,* will provide overwhelming testimony as to the systematic applications of foliar anatomy. Certainly no generic monograph can be said to be complete without studies on leaf anatomy.

REFERENCES

ARBER, AGNES, 1925. *Monocotyledons,* Cambridge: Cambridge University Press.
BAILEY, I. W., and NAST, CHARLOTTE G., 1944. "The comparative anatomy of the Winteraceae. V. Foliar epidermis and sclerenchyma," *Jour. Arnold Arb.,* **25:** 342–348.
CARLQUIST, S., 1957a. "The genus *Fitchia* (Compositae)," *Univ. Calif. Publ. Bot.,* **29:** 1–144.

CARLQUIST, S., 1957b. "Leaf anatomy and ontogeny in *Argyroxiphium* and *Wilkesia* (Compositae)," *Amer. Jour. Bot.*, **44**: 696–705.

———, 1958. "Anatomy of Guayana Mutisieae. Part II," *Mem. N. Y. Bot. Gard.*, **10**: 157–184.

———, 1959a. "Vegetative anatomy of *Dubautia, Argyroxiphium,* and *Wilkesia* (Compositae)," *Pac. Sci.*, **13**: 195–210.

———, 1959b. "Studies on Madinae: anatomy, cytology, and evolutionary relationships," *Aliso*, **4**: 171–236.

CHARTSCHENKO, W., 1932. "Verschiedene Typen des mechanischen Gewebes und des kristallinischen Ausbildungen als systematische Merkmale der Gattung *Allium*," *Beih. Bot. Centr.*, **50**(2): 183–206.

CHATIN, J., 1875. "Études histologiques et histogéniques sur les glandes foliaires intérieurs et quelques productions analogues," *Ann. Sci. Nat. Bot.*, sér. 6, **2**: 199–221.

COLOMB, G., 1887. "Recherches sur les stipules," *Ann. Sci. Nat. Bot.*, sér. 7, **6**: 1–76.

COSTATIN, J., 1886. "Études sur les feuilles des plantes aquatiques," *Ann. Sci. Nat. Bot.*, sér. 7, **3**: 94–162.

DE WET, J. M. J., 1954. "Leaf anatomy and phylogeny in the tribe Danthoniae," *Amer. Jour. Bot.*, **41**: 204–211.

EAMES, A. J., 1953. "Neglected morphology of the palm leaf," *Phytomorphology*, **3**: 172–189.

EDELHOFF, E., 1886. "Vergleichende Anatomie des Blattes der Familie der Olacineen," *Bot. Jahrb.*, **8**: 100–153.

ETTINGHAUSEN, C., 1861. *Die Blatt-Skelete der Dicotyledonen mit besonderer Rücksicht auf die Untersuchung und Bestimmung der fossilen Pflanzenreste.* Wien.

FAHN, A., 1954. "The anatomical structure of the Xanthorrhoeaceae Dumort," *Jour. Linn. Soc. Lond.*, **55**: 158–184.

FOSTER, A. S., 1950a. "Morphology and venation of the leaf in *Quiina acutangula* Ducke," *Amer. Jour. Bot.*, **37**: 159–171.

———, 1950b. "Venation and histology of the leaflets in *Touroulia guianensis* Aubl. and *Froesia tricarpa* Pires," *Amer. Jour. Bot.*, **37**: 848–862.

———, 1951. "Heterophylly and foliar venation in *Lacunaria*," *Bull. Torrey Bot. Club*, **78**: 382–400.

———, 1959. "The morphological and taxonomic significance of dichotomous venation in *Kingdonia uniflora* Balfour f. and W. W. Smith," *Notes Roy. Bot. Gard. Edinburgh*, **23**: 1–12.

GODFRIN, J., 1884. "Recherches sur l'anatomie comparée des cotylédons," *Ann. Sci. Nat. Bot.*, sér. 6, **19**: 5–158.

GUTTENBERG, H. VON, 1926. "Die Bewegungsgewebe," in K. Linsbauer, ed., *Handbuch der Pflanzenanatomie*, **5**(2) : v + 289.

HABERLANDT, G., 1894. "Ueber Bau und Funktion der Hydathoden," *Ber. Deutsch. Bot. Ges.*, **12**: 367–378.

HAGERUP, O., 1953. "The morphology and systematics of the leaves in Ericales," *Phytomorphology*, **3**: 459–464.

HUBBARD, C. E., 1948. "The genera of British grasses," in J. Hutchinson, *British flowering plants* (pp. 284–348).

KERNER, A. J., and OLIVER, F., 1895. *The natural history of plants*, Vol. I, New York: Holt, Rinehart and Winston.

KURT, J., 1929. "Über die Hydathoden der Saxifrageae," *Beih. Bot. Centr.*, 46(1): 203–246.

LECLERC DU SABLON, M., 1887. "Recherches sur l'enroulement des vrilles," *Ann. Sci. Nat. Bot.*, sér. 7, 5: 5–50.

LEINFELLNER, W., 1959. "Die falschen Rollblätter der Frankeniaceen, in Vergleich gesetzt mit jenen der Ericaceen," *Öst. Bot. Zeit.*, 106: 325–351.

McMINN, H. E., 1942. "A systematic study of the genus *Ceanothus*," in M. Van Rensselaer and H. E. McMinn, *Ceanothus*, Santa Barbara: Santa Barbara Botanic Garden.

METCALFE, C. R., 1956. "Some thoughts on the structure of bamboo leaves," *Bot. Mag.*, Tokyo, 69: 391–400.

————, and CHALK, L., 1950. *Anatomy of the dicotyledons*, Oxford: Clarendon Press.

MEYER, F., 1923. "Das trophische Parenchym. A. Assimilationsgewebe," in K. Linsbauer, ed., *Handbuch der Pflanzenanatomie*, 4(1): vii + 85.

MOENS, P., 1956. "Ontogenèse des vrilles et différentiation des ampoules adhésives chez quelques végétaux (*Ampelopsis, Bignonia, Glaziovia*)," *La Cellule*, 57: 371–401.

MORLEY, T., 1953a. "The genus *Mouriri* (Melastomaceae). A sectional revision based on anatomy and morphology," *Univ. Calif. Publ. Bot.*, 26: 223–312.

————, 1953b. "A new genus and three new species in the Memecyleae (Melastomaceae)," *Amer. Jour. Bot.*, 40: 248–255.

MOSER, H., 1934. "Untersuchungen über die Blattstruktur von *Atriplexarten* und ihre Beziehungen zur Systematik," *Beih. Bot. Centr.*, 52(2): 378–388.

MÜLLER, C., 1909. "Beiträge zur vergleichenden Anatomie der Blätter der Gattung *Agave* und ihre Verwertung für die Unterscheidung der Arten," *Bot. Zeit.*, 67: 93–139.

MÜLLER, E., 1944. "Die Nervatur der Nieder- und Hochblätter," *Bot. Archiv.*, 45: 1–92.

PFEIFFER, H., 1928. "Die pflanzlichen Trennungsgewebe," in K. Linsbauer, ed., *Handbuch der Pflanzenanatomie*, 5(3): viii + 236.

PRAY, T. R., 1955. "Foliar venation of angiosperms. IV. Histogenesis of the venation of *Hosta*," *Amer. Jour. Bot.*, 42: 698–706.

PRIEMER, F., 1893. "Die anatomischen Verhältnisse der Laubblätter der Ulmaceen (einschl. Celtideen) und die Beziehung zu ihrer Systematik," *Bot. Jahrb.*, 17: 419–475.

SAUVAGEAU, C., 1891. "Sur les feuilles des quelques monocotylédones aquatiques," *Ann. Sci. Nat. Bot.*, sér. 7, 13: 103–296.

SCHNEE, L., 1939. "Ranken und Dornen," in K. Linsbauer, ed., *Handbuch der Pflanzenanatomie,* 9: 1–24.

SCHUSTER, W., 1910. "Zur Kenntniss der Aderung des Monocotylenblattes," *Ber. Deutsch. Bot. Ges.*, 28: 268–278.

SMITHSON, E., 1956. "The comparative anatomy of the Flagellariaceae," *Kew Bull.*, (3): 491–501.

SOLEREDER, H., 1908. *Systematic anatomy of the dicotyledons* (trans. by Boodle and Fritsch), Oxford: Clarendon Press.

STEBBINS, G. L., 1956. "Cytogenetics and evolution of the grass family," *Amer. Jour. Bot.*, **43**: 890–905.

TAKHTAJAN, A. L., 1959. *Essays on the evolutionary morphology of plants* (trans. by Olga H. Gankin), Amer. Inst. Biolog. Sci., Washington, D. C.

TOMLINSON, P. B., 1956. "Studies in the systematic anatomy of the Zingiberaceae," *Jour. Linn. Soc. Lond.*, **55**: 547–592.

———, 1959. "An anatomical approach to the classification of the Musaceae," *Jour. Linn. Soc. Lond.*, **55**: 779–809.

TRAPP, F., 1933. "A study of the foliar endodermis in the Plantaginaceae," *Trans. Roy. Soc. Edinburgh*, **57**: 523–546.

VESQUE, J., 1883. "Contributions à l'histologie systématique de la feuille des Caryophyllinées," *Ann. Sci. Nat. Bot.*, sér. 6, **15**: 105–148.

WUNDERLICH, ROSALIE, 1950. "Die Agavaceae Hutchinsons im Lichte ihres Embryologie, ihres Gynözeums-, Staubblatt- und Blattbaues," *Öst. Bot. Zeit.*, **97**: 438–502.

ZALENSKI, V., 1902. "Über die Ausbildung der Nervation bei verscheidenen Pflanzen," *Ber. Deutsch. Bot. Ges.*, **20**: 433–440.

ZIMMERMANN, J. G., 1932. "Über die extrafloralen Nektarien der Angiospermen," *Beih. Bot. Centr.*, **49**(1): 99–196.

chapter ten ▶ Flower, and
Associated Topics

Floral anatomy, sometimes termed floral morphology, has for many botanists the connotation of vascular anatomy of flowers. Although studies of this nature do account for much work in floral anatomy, other aspects cannot be neglected, because they offer much information of value to comparative anatomy. Morphological studies of the flower in its broadest sense may be considered under the following headings:

1. *The Inflorescence:* Studies on inflorescences have occupied an awkward position between the fields of taxonomy and anatomy. Occasional forays into inflorescence morphology by studies of one or both of these disciplines have resulted in a few, but unfortunately few, excellent studies. From the taxonomist's point of view, Rickett (1955) has performed a useful service in delimiting and redefining inflorescence types, and in emphasizing that many types, such as umbels, are an unnatural grouping of diverse phenomena. The study of Mann (1959) on umbels of *Allium* is a model of the type of anatomical work needed to re-examine inflorescence types in the light of Rickett's commentary. Inflorescence studies are often not separable from studies on floral anatomy. For example, interpretation of flowers in catkinlike inflorescences requires comprehension both of floral units and of the inflorescence as a whole (Abbe, 1938; Abbe and Earle, 1940; Hjelmqvist, 1948). That studies of inflorescences may be useful

124

from a taxonomic point of view is shown by such studies as that of Mann and by Philipson's (1948) studies on Compositae and allied families. Additional information is provided by Eames (1961).

2. *Developmental Anatomy:* Studies pertaining to ontogenetic stages in development of floral parts may yield important dividends for comparative purposes. However, because the focus of these studies is often upon other problems, they are not considered here in detail.

3. *Vascular Anatomy:* Under this heading are included investigations that attempt to show the basic nature of floral vascularization and phylogenetic modifications of the vascular system.

4. *Floral Histology:* This category includes cell types and their distribution in the flower.

5. *Fruit Histology:* Studies in this area deal with cell types (including vascular tissue) and their distribution in the mature or maturing gynoecium and associated parts.

6. *Seed Histology:* This topic includes histological studies on the seed (especially seed coat) with reference to origin of layers of the seed coat from parts of the ovule.

7. *Embryology:* This term in its strictest sense would refer to the development and morphology of the embryo. Ironically, it has come to refer to those steps preceding embryo development— that is, megagametogenesis. These topics have become botanical microcosms with a complicated literature of their own and cannot be considered in detail here. The reader is referred in particular to Maheshwari's (1950) book and to three main reference works: Souèges (1934–1939) discusses many aspects of plant embryology; Schnarf (1929) covers material on embryo-sac development; and Johansen (1950) summarizes literature on development of the embryo itself. The usefulness of embryology to taxonomy is emphasized by Maheshwari (1950, Chapter 11) and Mauritzon (1939). Excellent examples of taxonomic use of embryological data have been offered by Cave (1953) and Maheshwari (1958).

8. *Palynology:* Like embryology, study of pollen has become a distinct subscience with its own methods and terminology. Literature in this rapidly developing field is beyond the scope of this book, and the reader is urged to consult the important

compendia and textbooks of Erdtman (1952), Faegri and Iversen (1950), and Wodehouse (1935). Pollen "atlases" such as those of Cranwell (1953) and Ikuse (1956) are worthy of mention. Recent papers dealing with terminology and means of presentation include those of Erdtman (1959) and Erdtman and Vishnu-Mittre (1958). Contributions that consider phylogeny of pollen-grain features include those of Pohl (1928) and Wodehouse (1928, 1936). Abundant reference to systematic value of pollen-grain morphology—a field of tremendous value—is offered by Erdtman (1952).

Topics 3, 4, 5, and 6 above are considered in detail in the following discussions.

VASCULAR ANATOMY

The usefulness of data from vascular anatomy is great, but the caution needed in interpretation cannot be overemphasized. Values in studies of floral vascularization depend at least in part on the assumption, tacit or implied, that the vascular pattern of a flower is "conservative"; that is, that it evolves at a slower rate than the gross appearance of the flower. This may be true, but exceptions do occur. Failure to attribute a degree of plasticity to vascular systems of flowers has resulted in questionable interpretations, such as those of Saunders (see Eames, 1931). A number of interesting studies of floral morphology are offered by Eames (1961).

The primitive flower and its modifications

Recent investigations of Ranales have added much to our understanding of the nature of floral vascularization. The chief items of interest are (1) nature of vascular supply—type of nodes and number of traces; (2) elaborateness of vascularization (particularly in stamens); (3) position of microsporangia and ovules in relation to the vein system; and (4) the beginning stages of alterations that result in syncarpy, cohesion, etc.

Sepals; tepals. As Eames (1931) states, sepals are most easily compared with leaves (or bracts) in their venation and other structural features. Sepals are predominantly three-trace structures, like many leaves (Fig. 10-1B). The bracts and tepals of *Austrobaileya* (Fig. 10-1A) show a double trace each, however. This condition corresponds to the unilacunar two-trace nodes of the vegetative shoot (see Chapter 6), which are regarded as primitive. Among relatively primitive flowers, similarity of leaf primordia to those of sepals and bracts (Tepfer, 1953) is especially noteworthy.

Fig. 10-1. Laid-out vascularization patterns of two ranalean flowers, to show a primitive vascular pattern (left) and a somewhat more advanced one (right). A, *Austrobaileya scandens;* B, *Ranunculus repens.* Note that in *Austrobaileya,* all traces other than carpel traces are of the dual-trace type, as in leaves (Fig. 6-3E). In *Ranunculus,* sepals are related to trilacunar nodes, petals are related to a tripartite trace from a single gap, and stamens and carpels are single-trace structures. B = bract; C = carpel; P = petal; S = stamen; Se = sepal; Sn = staminode; T = tepal. (A, redrawn from Bailey and Swamy, 1949; B, redrawn from Tepfer, 1953.)

Petals. In contrast with sepals, petals often appear to be one-trace structures (Fig. 10-1B), although the trace may ramify in the petal (Eames, 1931). Petals may, in some instances, be related to more than one trace (for example, Winteraceae; see Bailey and Nast, 1944). The theory that petals are derived from stamens (Eames, 1931) was suggested on account of the predominantly one-trace nature of both petals and stamens in angiosperms, in contrast with sepals and carpels. This interpretation has found support from such workers as Moseley (1958), from whose work in Nymphaeaceae a series of stages (Fig. 10-2) is represented here. In primitive ranalean groups, such as Winteraceae, where three-trace stamens occur, the occurrence of three-trace petals seems noteworthy (Bailey and Nast, 1944). On the basis of ontogenetic studies, Tepfer (1953) finds great similarity between petals and stamens, and finds staminal origin of petals tenable.

Stamens. As noted early by Eames (1931), stamens of most angiosperms are one-trace structures, but among Ranales, stamens with three independent traces occur (Figs. 10-2, 10-3). Staminal

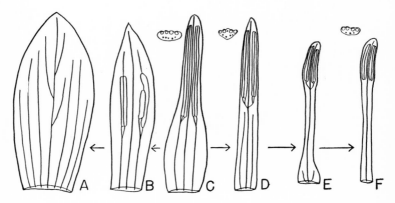

Fig. 10-2. Trends of stamen phylogeny in Nymphaeaceae that have a common ancestral type (C, a flat, three-trace stamen with adaxial sporangia) but that result in two different products, a petal (A) in one case and a filiform stamen (F) in the other. A, petal of *Nymphaea Heudelottii;* B, outer stamen of *Nymphaea Heudelottii;* C, central stamen of *Nymphaea odorata;* D, E, F, inner stamens of *Nymphaea odorata.* (Redrawn from Moseley, 1958.)

structure of *Umbellularia* (Kasapligil, 1951) is similar to that of some Monimiaceae (Money, Bailey, and Swamy, 1950) and suggests affinity between these families. The elongate stamens, with three, often branched, traces, ally the families Himantandraceae, Degeneriaceae, and Magnoliaceae (Fig. 10-3). The basic pattern in Nymphaeaceae (Fig. 10-2) appears to be a three-trace structure. The contrast between these conditions and two-trace conditions may seem significant, but the difference between a unilacunar two-trace node and a trilacunar node is probably not very great (see Chapter 6). Transitional conditions may be present in stamens of *Sarcandra, Austrobaileya,* and *Umbellularia* (Fig. 10-3), and a reduced two-trace condition is probably present in the vesselless genus *Amborella*

(Bailey and Swamy, 1948). A feature worthy of mention in several of the above cases is the tendency of primitive stamens to be flat and leaflike, with sporangia embedded in the surface rather than

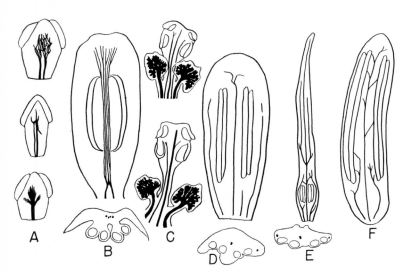

Fig. 10-3. Stamens of various ranalean species, showing primitive types of construction. A, *Sarcandra glabra* (Chloranthaceae), three stamens that show expressions of a dual-trace nature and (below) a trifid single trace. B, *Austrobaileya scandens* (Austrobaileyaceae), showing dual-trace nature of supply, with forking of veins above; note flat, leaflike shape (transection shown below, adaxial face below). C, *Umbellularia californica* (Lauraceae), two stamens, showing dual-trace (above) and three-trace (below) conditions; heavily vascularized structures at right and left in each are "staminal glands." D, *Degeneria vitiensis* (Degeneriaceae), showing flat, leaflike stamen with three main veins (transection below, abaxial face downward). E, *Galbulimima* (*Himantandra*) *Belgraveana* (Himantandraceae), long, sporophyll-like stamen with three main traces and much branching of veins (below, transection with abaxial face downward). F, *Magnolia stellata* (Magnoliaceae), showing three main veins, each of which is branched (sporangia are on adaxial surface). Magnifications various. (All redrawn: A, from Swamy and Bailey, 1950; B, from Bailey and Swamy, 1949; C, from Kasapligil, 1951; D, from Bailey and Smith, 1942; E, from Bailey, Nast and Smith, 1943; F, from Canright, 1952.)

borne along the margins or at the apex. Ontogenetically, the difference between one-trace stamens and leaves appears to be the failure of the two lateral strands of procambium to develop (Tepfer, 1953);

early stages appear quite similar. For phylogenetic interpretations of leaflike and stipulate stamens, see Leinfellner (1956a, 1956b).

Carpels. Eames (1931) visualized the carpel as primitively a three-trace structure. These three traces are the dorsal (abaxial), or

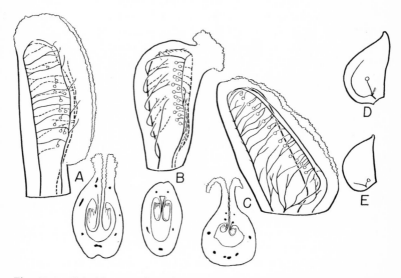

Fig. 10-4. Primitive carpels and stages in carpel evolution in Ranales. Veins on nearer face of carpel shown unbroken, veins on farther face of carpel shown in broken lines; circles represent ovules; only half of the veins shown in carpel of C, above. A, *Drimys* (sect. *Tasmannia*) *piperita*; B, *Drimys* (sect. *Drimys*) *Winteri*; C, *Degeneria vitiensis*; D, *Ranunculus repens*; E, *Ranunculus aquatilis*. Carpel evolution in the genus *Drimys* has involved reduction of stigmatic surfaces and loss of veins running into the stigmatic margins. *Degeneria* shows an open-carpel type reflecting a primitive type of construction. In *Ranunculus*, carpels show a highly reduced version of the types shown in A-C. E shows loss of adaxial veins, shortening of abaxial vein. Magnifications various. (All redrawn: A, B, from Bailey and Nast, 1943; C, from Swamy, 1949; D, from Tepfer, 1953; E, from Eames, 1931.)

midvein, and the two lateral (adaxial), or ventral traces, from which traces to the ovules branch. Study of ranalean carpels (Fig. 10-4) has modified these concepts somewhat. Bailey and Swamy (1951) claim that (1) ovules are not truly marginal, primitively, but are borne on the adaxial surface at a distance from the margins (for example, Fig. 10-4A). Because relatively few angiosperms show this condition, alteration to a marginal position concomitant with loss of stigmatic carpel edges (Fig. 10-4B) must have been rapid. Position of ovules on the adaxial surface of a carpel might be enhanced in significance if a similar position is primitive for microsporangia (see Fig. 10-3). (2) Carpels are primitively conduplicate rather than cylindrical (see Fig. 10-4A, C, below). This feature is related to (3) margins of carpels are envisioned as stigmatic areas primitively; this is correlated with the primitive open nature of carpels in *Drimys* (Fig. 10-4A), *Degeneria* (10-4C), and *Galbulimina* (*Himantandra*). Most carpels of angiosperms show a more advanced condition, in which the margins of the carpels are either fused or lost. Bailey and Swamy postulate that stigmatic areas have become restricted to the apical portion of the carpels (Fig. 10-4B); prolongation of the carpel tip into a style, perhaps to enhance pollination mechanisms, may have followed. Ontogenetic considerations appear to support these concepts. The work of Periasamy and Swamy (1956) showed that in *Cananga,* the carpel is conduplicate, not inrolled, and the ovules are borne on traces related to the abaxial (dorsal) rather than to adaxial veins.

The three-trace nature of the primitive carpel seems clear, although additional traces may be present, as in *Austrobaileya* (Fig. 10-1A) or *Degeneria* (Fig. 10-4C). The double trace of the midvein (Fig. 10-1A) may be significant as it is in nodal considerations of other organs in primitive angiosperms, and Eames (personal communication) suggests that a double carpel midvein is not as uncommon among angiosperms as might be thought on the basis of existing descriptions. Reduction of the three main traces of the carpel have taken various forms, both in apocarpous and syncarpous forms; a reduction series in an apocarpous genus is shown in Fig. 10-4D-E.

Bailey and Swamy (1951) have further explored several possible origins of syncarpous conditions. From the apocarpous condition (Fig. 10-5A) they believe that syncarpy may have been derived in three ways, which are illustrated and described in Fig. 10-5B, C, and D, respectively. The evolution of various placentation types, their distribution and terminology, are reviewed in a paper by Puri (1952).

Supracarpellary vascularization. Because he based his anatomical interpretations of the flower on the theory that it is a truncated shoot, Eames (1931) sought "vestigial" vascularization above the point in the flower where the uppermost carpellary traces depart. He reports several instances (1931), such as *Aquilegia,* in which phloem cells are present in such a position (see Tepfer, 1953, for a discussion). In flowers where such "vestigial" traces might be

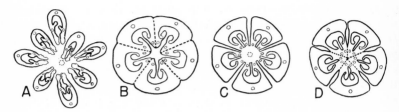

Fig. 10-5. Diagrammatic transectional views of (A), a spiral apocarpous gynoecium with conduplicate carpels, each with an abaxial and a pair of submarginal adaxial veins; and (B-D) syncarpous tendencies. B shows lateral concrescence of a whorl of carpels; C shows adnation of the free margins of conduplicate carpels to the torus; D shows concrescence of adaxial margins of carpels. (B-D, redrawn from Bailey and Swamy, 1951.)

expected to occur, they have not been found (Bailey and Nast, 1944). Such vascularization might merely represent traces to carpels that aborted ontogenetically. In any case, its absence does not negate the determinate-shoot nature of the floral axis, for developmental anatomy has shown that traces almost always originate in relation to actual appendages.

The inferior ovary

Until recently, botanical opinion was divided on the mode of origin of the inferior ovary. Studies on floral anatomy have shown that many plants with inferior ovaries have probably achieved this condition by the appendicular method—that is, by progressive adnation of the three outer whorls of the flower, or a cup formed by adhesion of these whorls, to the gynoecium. Douglas (1944) cited a number of instances of such origin in her first review on this topic, and in a recent review (1957) has cited other cases. Vascular patterns may be used to demonstrate the appendicular type of origin, as Eames (1931) has done in Ericales, shown in Fig. 10-6. The fact, in this series of cases, that fusion of veins lags behind fusion of parts enables us to interpret the origin of the ovary quite readily, but more important is the presence of the various intermediate stages. An additional case of evolution in an inferior ovary of the appendicular type is described below in connection with Compositae.

The other possible mode of origin, sometimes known as the cauline or axial cup type, involves sinking of the gynoecium into the floral axis (receptacle). This type did not win general recogni-

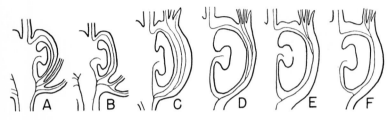

Fig. 10-6. Evolution of an inferior ovary in Ericales by means of adnation of whorls. Half of a longitudinal section of a flower is shown in each. A, *Pyrola secunda,* with no fusion of organs; B, *Andromeda glaucophylla,* stamen base adnate to corolla, traces to these organs fused nearly to base of stamen; C, *Gaylussacia frondosa,* showing union of parts with traces remaining largely free from each other; D-F, *Vaccinium* spp., showing progressive union of traces in the inferior ovary of this genus; D, *V. corymbosum;* E, *V. canadense;* F, *V. macrocarpon.* (Redrawn from Eames, 1931.)

tion until a number of cases (reviewed by Douglas, 1957) had been documented. Smith and Smith (1942) showed that in *Darbya,* as well as in other members of Santalaceae (1943), this type could be identified by the fact that bundles supplying the ovary run downward from the axis tip to the base of the ovary and are therefore inverted (phloem adaxial) in orientation. This criterion, where applied to other groups, shows that the axial cup mode of origin has been followed independently in several families. Although the opening stages of the sinking of the gynoecium into the axis are not demonstrated by cacti, the progressively greater sinking, and modification of, the gynoecium are visible in a series of cases (Fig. 10-7) described by Tiagi (1955). The fact that the genera selected form a sequence of specialization—suggested previously on other grounds—makes this series particularly compelling. The occurrence of an inferior ovary interpreted as of the cauline type in *Mesembryanthemum* of the

Aizoaceae, a family supposedly related to Cactaceae, is of considerable phylogenetic interest (Ihlenfeld, 1960).

The ontogenetic processes by which an inferior ovary is achieved in any species need not be expected to reflect its phylogeny

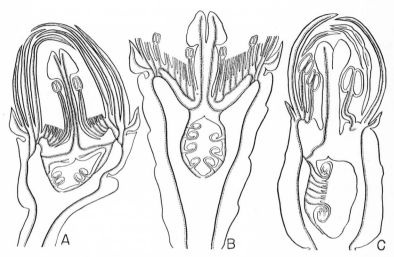

Fig. 10-7. Stages in evolution of the inferior ovary of Cactaceae, showing less specialized (A) and progressively more specialized (B, C) instances of sinking of gynoecium into the receptacle. Xylem indicated by unbroken lines, phloem by dotted lines. Presence of inverted bundles in gynoecium is considered evidence of the cauline nature of the inferior ovary. A, *Pereskia bleo;* B, *Opuntia Dillenii;* C, *Rhipsalis cassutha.* (Redrawn from Tiagi, 1955.)

directly. The writer does not believe that skepticism toward vascular anatomy by workers whose primary concern is with ontogenetic phenomena in flowers is wholly justified.

Usefulness of vascular anatomy

In addition to derivation of sequences such as those mentioned above, vascular anatomy of the flower as a whole can be applied to studies on the phylogeny of a particular group. The review of Eames (1953) gives a number of such examples. The body of literature on floral vascularization is tremendous. Such bibliographies as that of Rao (1951) may help the student, but there is no single bibliographic tool (other than abstracting journals) for finding works on these topics.

Ideas on floral venation of Compositae and its evolution are summarized in Fig. 10-8. These form a convenient example because the most primitive flowers of Compositae show a venation pattern that is simple compared with that of most angiosperms and may be represented in its entirety with relative ease. Most of the species illustrated here as examples of particular trends have been discussed earlier (Carlquist, 1957a, 1957b; Stebbins, 1940). Although a few members of tribes other than those for which drawings are shown may retain primitive venation patterns (for example, *Proteopsis* of the Vernonieae), by far the majority show venation of the sort termed "simplified type" here.

Trends in vascular evolution of disc flowers of Compositae are based upon the following interpretations:

1. *Achene and Style:*

(a) Two main lines have developed from an ancestral type. These types may be termed helianthoid and mutisioid respectively. Primitive members of the mutisioid line have retained ancestral patterns of achene venation to a much greater extent than those in the helianthoid line, but some features, such as vascularized calyx (=pappus) have been retained to a greater extent in the helianthoid line. Presumably the hypothetical ancestral type had a vascularized calyx, perhaps composed of five lobes.

(b) Ancestrally (as in *Stenopadus*) there is in the achene wall (wall of the inferior ovary) an outer circle of 10 bundles, representing the traces to 10 corolla bundles and the five stamen bundles united to five of the corolla bundles. An inner circle of bundles, representing gynoecium bundles, is also present ancestrally. Conformation of the inner circle of bundles often suggests (corresponding to the bifid style) a bicarpellate gynoecium, in which the dorsal traces would be to the left and right, inner circle, in the transection of the *Stenopadus* achene in Fig. 10-8. Occasional reports of bicarpellate ovaries in Compositae may provide additional evidence in this regard. "Dorsal" traces of the inner circle extend upward as the two main style bundles. In the more primitive Mutisieae, the "lateral" carpellary traces

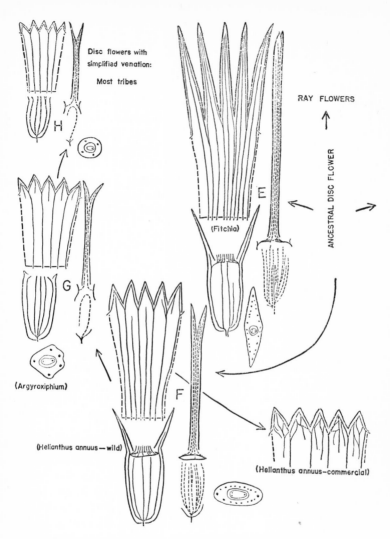

Disc flowers with
simplified venation:

Most tribes

H

RAY FLOWERS

ANCESTRAL DISC FLOWER

E

(Fitchia)

G

(Argyroxiphium)

F

(Helianthus annuus — wild)

(Helianthus annuus—commercial)

Fig. 10-8. Selected trends of evolution in venation and form of flowers of Compositae, with indications of systematic distribution of the types. Although venation is figured for particular genera, the instances figured do not in all respects correspond to a desired hypothetical type (for example, B would have median veins in corolla lobes if the phylogenetic series were perfect). Literal derivation (for example, *Dubyaea* of Cichorieae from *Glossarion* of Mutisieae) is not intended. Corolla is shown laid out, the "cut" indicated by broken lines. Ovary below corolla, showing outer

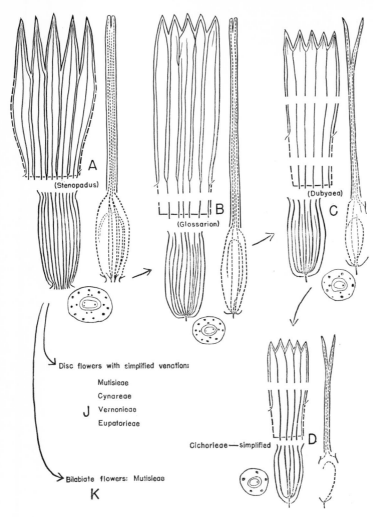

series of ovary veins; at right in each case, the gynoecial unit (broken lines) consisting of stigmatic branch traces, inner series of ovary bundles, and ovule trace or traces, as well as portions (at base) of outer series of bundles with which gynoecial bundles are connected. All ovaries seen from a lateral view, except for E, in which ovary is seen in dorsiventral view; ovary transection in each shown with adaxial surface at left. Further explanation in text.

are connected upwardly with a lateral pair of style traces, so that a total of four traces is present in the style of such species. Because a number of species with achene and style vascularization of a primitive type have a dichotomous ovule trace, ovule vascularization may be worthy of consideration in this regard also.

(c) In the mutisioid line, simplifications in ovary structure have resulted in the loss of gynoecium bundles other than the ovule trace, loss of the lateral pair of style bundles, and reduction of the outer achene wall bundles from 10 to five. Apparently the dorsiventral gynoecial traces have been united with nearby outer achene wall bundles before the lateral gynoecial traces were united with outer bundles or lost. Loss of outer achene-wall bundles has lagged behind loss of median corolla bundles (the bundles median in each corolla lobe). The five outer bundles connected with median corolla-lobe bundles are lost following the disappearance of gynoecial traces (for stage in loss, see drawing of *Dubyaea* in Fig. 10-8).

(d) In the helianthoid line, multiplication of bundles has probably taken place in *Fitchia* and *Helianthus,* perhaps in response to increase in achene size. That this is likely is suggested by the elaborate venation in the commercial sunflower both in achene and corolla (see Fig. 10-8, left, F). Union of inner (carpellary) bundles to outer achene-wall bundles—or their change to an outer position—must have taken place early, because no Heliantheae have been shown to have an inner and outer series of bundles. As in the mutisioid line, however, four style bundles are present in some genera. These are reduced to two in advanced Heliantheae and allied tribes. Likewise, the number of achene-wall bundles is reduced to five or four (or even two, in some reduced species) from the ten that may be assumed to have been present primitively.

(e) Noting that the primitive flower in angiosperms at large has a vascular cylinder, the reader may wonder how Compositae reflect a cylindrical vascular supply. Only a single vascular bundle supplies the base of the achene in most Compositae. However, as shown in Fig. 10-8 for *Stenopadus,* a cylinder of bundles is primitively present, and evolutionary reduction to the single bundle has taken place early in the family.

2. *Corolla and Stamens:*

(a) **Primitively**, corolla venation consisted of one median bundle and two lateral bundles in each lobe. Sinuses in primitive Compositae were probably deep, perhaps nearly separating the lobes (suggested by deep sinuses in primitive Mutisieae and lack of union between adjacent lateral veins). Stamen traces are associated with each pair of lateral bundles (stamen traces not shown on Fig. 10-8). Stamen traces are reduced, in the most advanced Compositae, to bundles of phloem alone. The three veins at the tip of each corolla lobe may or may not be fused; freely terminating veins characterize lobe tips in many primitive Compositae. In addition to the median and lateral veins in each corolla lobe, other veins may be present; these probably represent (in most instances, at least) a secondary increase in vascularization.

(b) In the mutisioid line, adjacent lateral bundles have become united into pairs. Following this simplification, median bundles have been lost, wholly or in part, with the result that only the fused laterals are present. Fragmentary presence of median bundles is shown by *Dubyaea atropurpurea* (tips of lobes). Concomitant with, or following this loss of bundles, lobes have become united, a fact that is related in the venation pattern (see *Glossarion*). Ligulate forms have been evolved in Cichorieae and Mutisieae; the bilabiate form has been achieved in Mutisieae.

(c) In the helianthoid line, the primitive corolla is best represented by *Fitchia*. This corolla is essentially identical with that of *Stenopadus*. Union of adjacent laterals has progressed rather far in *Helianthus*, and only terminal vestiges of the median veins are present. *Argyroxiphium* represents a case similar to *Dubyaea*, in which very short vestiges of median bundles occur in lobe tips. The simplified corolla that has resulted from this reduction series is the same as that in actinomorphic disc flowers of most members of the mutisioid line. Corolla vascularization increase in the commercial *Helianthus annuus* probably correlates with greater size of flowers.

3. *General Conclusions:* Because the most primitive Heliantheae and Mutisieae lack ray flowers, and because ray flowers, like ligulate flowers, are modifications of disc flowers, the family prob-

ably lacked ray flowers primitively. The ray flower is unlike the ligulate flower in that there are three lobes (usually) evident at the tip of the ray. The other two lobes have probably been incorporated into the margins of the expanded portion of the ray, as venation patterns indicate. Ray flowers of many Compositae show increase, rather than decrease, in number of veins, as compared with a primitive disc flower.

The interpretations given above were developed largely from anatomical data, but correlations with data in the taxonomic system were of value in reaching these interpretations, and no type of evidence was overlooked.

FLORAL HISTOLOGY

Many flowers are composed not merely of unvaried parenchyma but contain many details of construction and histology worthy of consideration.

Presence and variety of sclerified cells and trichomes at the tips of corolla lobes have been used as specific criteria in *Fitchia* (Carlquist, 1957a) and *Calycadenia* (Carlquist, 1959). Morley (1953) has shown the taxonomic value of sclereids in the calyx and hypanthium of *Mouriri*. Histology of flower-opening mechanisms in valvate floral organs is reviewed by Sigmond (1929). Tanniniferous cells (Jonesco, 1932) or oil cells are often characteristic of certain flowers; various types of pigmentation offer diagnostic criteria, as outlined by Möbius (1927).

Papers with much important comparative data on floral nectaries include those of Fahn (1953) on morphology, Feldhofer (1932) on histology, and Frei (1955) on venation. Of considerable interest is Brown's (1937) paper relating nectary morphology to angiosperm phylogeny.

Anatomy of anthers, including patterns of endothecial thickenings, and contractile anthers are considered by Leclerc du Sablon (1885); these thickenings, as well as histology of contractile anthers, are reviewed by Guttenberg (1926). For studies of poricidal anthers, see Matthews and MacLachlan (1929).

Anatomy of styles and stigmas has been explored by Guéguen (1901–1902). For information on sensitive styles and stigmas, see Guttenberg (1926).

FRUIT HISTOLOGY

Studies on the fruit are logically a continuation of those dealing with venation and histology of the flower. Studies admirable for this approach include those of Tepfer (1953) on *Aquilegia* and Swamy (1949) on *Degeneria*. Descriptions of this nature should mention from what portions of the ovary distinctive layers or cell types arise.

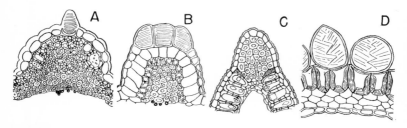

Fig. 10-9. Transection of achene (fruit) wall of species of Anthemideae (Compositae), showing distinctions in structure corresponding to the taxonomic system. A, *Anthemis peregrina;* B, *A. chia;* C, *A. tinctoria;* D, *Ormenis fuscata*. Ridge of fruit shown in A-C; epidermal cells containing lines are slime-producing cells. All ×165. (Redrawn from Briquet, 1916.)

Descriptions of the histology of the fruit wall, sometimes called "carpology," are scattered through the literature of various families. Such literature is often also concerned with seed development and embryology, and may be located under these topics. Fleshy fruits are reviewed by Garcin (1890) and dry-pericarp fruits by Krauss (1866–67). Dehiscence mechanisms are discussed by Guttenberg (1926). Information on the nature of the fruit wall can be invaluable; for example, much of the systematics of Umbelliferae rests upon it. Fruit morphology and anatomy have much to offer in aiding systems of fruit terminology, but such new systems (for example, Winkler, 1940) may not find ready acceptance.

Although "carpological" studies have recently been neglected while other areas of plant anatomy have developed in greater measure, there can be no question as to the value of such data. For example, the figures of achene transections of Anthemideae reproduced here (Fig. 10-9) from the work of Briquet (1916) show interesting correlations with the taxonomic system. Many more examples could have been selected, and one has only to imagine the tremendous variety in gross morphology and texture of fruits to visualize the wealth of anatomical diversity in these structures.

SEED HISTOLOGY

Like studies of the pericarp, comparative studies of seed anatomy are mostly concerned with contents of cells, cell types and distribution, and origin of layers in terms of the ovule. Most studies are closely related to anatomy of the embryo and endosperm, and may be found in the literature on plant embryology. Although not recent, the extensive compilation of Netolitzky (1926), arranged systematically, is an admirable review, basic to this field. Much

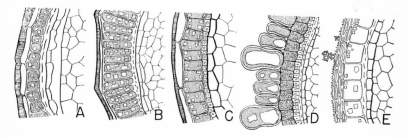

Fig. 10-10. Sections of seed coat of species of Papaveraceae, showing distinctions in structure. Note variations in cuticle, width of epidermal and hypodermal cells; contents (stippled) and crystals, and presence or absence of subhypodermal sclereids. A, *Chelidonium majus;* B, *Glaucium flavum;* C, *Bocconia microcarpa;* D, *Romneya trichocalyx;* E, *Eschscholzia californica.* (Redrawn from Röder, 1958.)

information on seed anatomy can be found in a series of articles edited by B. G. Alexandrov (1950–1958) and in a book by Tsinger (1958).

To illustrate how data on seed anatomy may be used, the examples shown in Fig. 10-10 from the work of Röder (1958) are offered. Although these show striking generic differentiation, an underlying similarity could be used to define a basic condition, and evolutionary tendencies, within Papaveraceae. Likewise, the study of Singh (1953) on Cucurbitaceae deserves special mention. The monograph of Le Monnier (1872) is valuable for studies concerned with ovule venation.

A significant study with regard to evolutionary interpretations of embryo shape, relation to endosperm, and seed morphology has been offered by Martin (1946). Reeder's (1957) study of embryo morphology in relation to taxonomy merits attention. Comparative data on arillate seeds are given by Corner (1953), although few will agree with the theory that prompted this compilation. Finally, the comprehensive work of Wunderlich (1959) on endosperm of angiosperms opens interesting avenues of systematic and phylogenetic interpretation.

REFERENCES

ABBE, E. C., 1938. "Studies in the phylogeny of the Betulaceae. II. Extremes in the range of variation of floral and inflorescence morphology," *Bot. Gaz.*, **99**: 431–469.

———, and EARLE, T. T., 1940. "Inflorescence, floral anatomy and morphology of *Leitneria floridana*," *Bull. Torrey Bot. Club*, **67**: 173–193.

ALEXANDROV, B. G., ed., 1950–1958. "Morphologia i anatomia rastenii," Trudy Botanicheskogo Instituta im B. L. Komarova, 4 vols., Academii Nauk SSSR.

BAILEY, I. W., and NAST, CHARLOTTE G., 1943. "The comparative morphology of the Winteraceae. II. Carpels," *Jour. Arnold Arb.*, **24**: 472–481.

———, ———, 1944. "The comparative morphology of the Winteraceae. VI. Vascular anatomy of the flowering shoot," *Jour. Arnold Arb.*, **25**: 454–466.

———, ———, and SMITH, A. C., 1943. "The family Himantandraceae," *Jour. Arnold Arb.*, **24**: 190–206.

———, and SMITH, A. C., 1942. "Degeneriaceae, a new family of flowering plants from Fiji," *Jour. Arnold Arb.*, **23**: 356–365.

———, and SWAMY, B. G. L., 1948. "*Amborella trichopoda* Baill., a new morphological type of vesselless dicotyledon," *Jour. Arnold Arb.*, **29**: 245–254.

———, ———, 1949. "The morphology and relationships of *Austrobaileya*," *Jour. Arnold Arb.*, **30**: 211–226.

———, ———, 1950. "*Sarcandra*, a vesselless genus of the Chloranthaceae," *Jour. Arnold Arb.*, **31**: 117–129.

———, ———, 1951. "The conduplicate carpel of dicotyledons and its initial trends of specialization," *Amer. Jour. Bot.*, **38**: 373–379.

BRIQUET, J., 1916. "Études carpologiques sur les genres de composées *Anthemis*, *Ormenis*, et *Santolina*," *Ann. Conserv. et Jard. Bot. Genève*, **18–19**: 257–313.

BROWN, W. H., 1937. "The bearing of nectaries on the phylogeny of flowering plants," *Proc. Amer. Phil. Soc.*, **79**: 549–595.

CANRIGHT, J. E., 1952. "The comparative morphology and relationships of the

Magnoliaceae. I. Trends of specialization in the stamens," *Amer. Jour. Bot.,* **39:** 484–497.

CARLQUIST, S., 1957a. "The genus *Fitchia* (Compositae)," *Univ. Calif. Publ. Bot.,* **29:** 1–144.

———, 1957b. "Anatomy of Guayana Mutisieae," *Mem. N. Y. Bot. Gard.,* **9:** 441–476.

———, 1959. "Studies on Madinae: anatomy, cytology, and evolutionary relationships," *Aliso,* **4:** 171–236.

CAVE, MARION S., 1953. "Cytology and embryology in the delimitation of genera," *Chronica Botanica,* **14:** 140–153.

CORNER, E. J. H., 1953. "The durian theory extended. I," *Phytomorphology,* **3:** 465–476.

CRANWELL, LUCY M., 1953. "New Zealand pollen studies. The monocotyledons," *Bull. Auckland Inst. and Mus.,* 3, Cambridge, Mass.: Harvard University Press.

DOUGLAS, GERTRUDE E., 1944. "The inferior ovary," *Bot. Rev.,* **10:** 125–186.

———, 1957. "The inferior ovary. II," *Bot. Rev.,* **23:** 1–46.

EAMES, A. J., 1931. "The vascular anatomy of the flower with refutation of the theory of carpel polymorphism," *Amer. Jour. Bot.,* **18:** 147–188.

———, 1953. "Floral anatomy as an aid in generic limitation," *Chronica Botanica,* **14:** 126–132.

———, 1961. *Morphology of the angiosperms,* New York: McGraw-Hill.

ERDTMAN, G., 1952. *Pollen morphology and plant taxonomy. Angiosperms,* Stockholm: Almqvist and Wiksell.

———, 1959. "Some remarks on pollen and spore illustrations," *Pollen et Spores,* **1:** 15–18.

———, and VISHNU-MITTRE, 1958. "On terminology in pollen and spore morphology," *Grana Palyn.* (n.s.), **1:** 6–9.

FAEGRI, K., and IVERSEN, J., 1950. *Textbook of modern pollen analysis,* Copenhagen: Einar Munksgaard.

FAHN, A., 1953. "The topography of the nectary in the flower and its phylogenetical trend," *Phytomorphology,* **3:** 424–426.

FELDHOFER, E., 1932. "Beiträge zur physiologischer Anatomie der nuptalien Nektarien aus den Reihen der Dikotylen," *Beih. Bot. Centr.,* **50**(1): 459–670.

FREI, EVA, 1955. "Die Innervierung des floralen Necktarien dikotylen Pflanzenfamilien," *Ber. Schweiz. Bot. Ges.,* **65:** 60–114.

GARCIN, A.-G., 1890. "Recherches sur l'histogénèse des péricarpes charnus," *Ann. Sci. Nat. Bot.,* sér. 7, **12:** 175–401.

GUÉGUEN, F., 1901–1902. "Anatomie comparée du tissu conducteur du style et du stigmate des phanérogames," *Jour. de Bot.,* **15:** 265–300; **16:** 15–30 *et seq.*

GUTTENBERG, H. VON, 1926. "Die Bewegungsgewebe," in K. Linsbauer, ed., *Handbuch der Pflanzenanatomie,* **5**(2): v + 289.

HJELMQVIST, 1948. "Studies on the floral morphology and phylogeny of the Amentiferae," *Bot. Notiser suppl.,* **2**(1): 1–171.

IHLENFELD, H.-D., 1960. "Entwicklungsgeschichte, morphologische und systematische Untersuchungen an Mesembryathemen," *Feddes Rep.,* **63** (1): 1–104.

IKUSE, M., 1956. *Pollen grains of Japan,* Tokyo.

JOHANSEN, D. A., 1950. *Plant embryology*, Waltham, Mass.: Chronica Botanica.

JONESCO, S., 1932. "Sur la présence des tannins chez les fleurs," *Ann. Sci. Nat. Bot.*, sér. 10, **13**: 327–344.

KASAPLIGIL, B., 1951. "Morphological and ontogenetic studies of *Umbellularia californica* Nutt. and *Laurus nobilis* L.," *Univ. Calif. Publ. Bot.*, **25**: 115–240.

KRAUSS, G., 1866–67. "Ueber den Bau trockner Pericarpien," *Jahrb. Wiss. Bot.*, **5**: 83–126.

LECLERC DU SABLON, M., 1885. "Recherches sur la structure et la déhiscence des anthères," *Ann. Sci. Nat. Bot.*, sér. 7, **1**: 97–134.

LEINFELLNER, W., 1956a. "Medianstipulierte Staubblätter," *Öst. Bot. Zeit.*, **103**: 24–43.

————, 1956b. "Die blattartig flachen Staubblätter und ihre gestaltlichen Beziehung zum Bautypus des Angiosperm-Staubblattes," *Öst. Bot. Zeit.*, **103**: 247–290.

LE MONNIER, M. G., 1872. "Recherches sur la nervation de la graine," *Ann. Sci. Nat. Bot.*, sér. 5, **16**: 233–305.

MAHESHWARI, P., 1950. *An introduction to the embryology of the angiosperms*, New York: McGraw-Hill.

————, 1958. "Embryology and taxonomy," *Mem. Indian Bot. Soc.*, **1**: 1–9.

MANN, L. K., 1959. "The *Allium* inflorescence: some species of the section *Molium*," *Amer. Jour. Bot.*, **46**: 730–739.

MARTIN, A. C., 1946. "The comparative internal morphology of seeds," *Amer. Midl. Nat.*, **36**: 513–660.

MATTHEWS, J. R., and MacLACHLAN, C. M., 1929. "The structure of certain poricidal anthers," *Trans. Bot. Soc. Edinburgh*, **30**: 104–122.

MAURITZON, J., 1939. "Die Bedeutung der embryologischen Forschung für das natürliche System der Pflanzen," *Lunds Univ. Årsskr. N. F. Avd. 2*, **35**(15): 1–69.

MÖBIUS, M., 1927. "Die Farbstoffe der Pflanzen," in K. Linsbauer, ed., *Handbuch der Pflanzenanatomie* 3(1/1): vii + 200.

MONEY, LILLIAN L., BAILEY, I. W., and SWAMY, B. G. L., 1950. "The morphology and relationships of the Monimiaceae," *Jour. Arnold Arb.*, **31**: 372–404.

MORLEY, T., 1953. "The genus *Mouriri* (Melastomaceae)," *Univ. Calif. Publ. Bot.*, **26**: 223–312.

MOSELEY, M. F., 1958. "Morphological studies on the Nymphaeaceae. I. The nature of the stamens," *Phytomorphology*, **8**: 1–29.

NETOLITZKY, F., 1926. "Anatomie der Angiospermen-Samen," in K. Linsbauer, ed., *Handbuch der Pflanzenanatomie*, **10**(1): iv + 364.

PERIASAMY, K., and SWAMY, B. G. L., 1956. "The conduplicate carpel of *Cananga odorata*," *Jour. Arnold Arb.*, **37**: 366–372.

PHILIPSON, W. R., 1948. "Studies in the development of the inflorescence. V. The raceme of *Lobelia dortmanna* L. and other campanulaceous inflorescences," *Ann. Bot.*, n.s., **12**: 147–156.

POHL, F., 1928. "Die einfaltige Pollen, seine Verbreitung und phylogenetische Bedeutung," *Beih. Bot. Centr.*, **45**(1): 50–73.

PURI, V., 1952. "Placentation in angiosperms," *Bot. Rev.*, **18**: 603–651.

RAO, V. S., 1951. "The vascular anatomy of flowers. A bibliography," *Jour. Univ. Bombay*, **29**(5): 38–63.

REEDER, JOHN R., 1957. "The embryo in grass systematics," *Amer. Jour. Bot.*, **44**: 756–768.

RICKETT, H. W., 1955. "Materials for a dictionary of botanical terms. III. Inflorescences," *Bull. Torrey Bot. Club*, **82**: 419–445.

RÖDER, INGEBORG, 1958. "Anatomische und fluoreszensoptische Untersuchungen an Samen von Papaveraceen," *Öst. Bot. Zeit.*, **104**: 370–381.

SCHNARF, K., 1929. "Embryologie der Angiospermen," in K. Linsbauer, ed., *Handbuch der Pflanzenanatomie*, **10**(2): xi + 689.

SIGMOND, H., 1929. "Vergleichende Untersuchungen über die Anatomie und Morphologie von Blütenknospenverschlüssen," *Beih. Bot. Centr.*, **46**(1): 1–67.

SINGH, B., 1953. "Studies on the structure and development of seeds of Cucurbitaceae," *Phytomorphology*, **3**: 224–239.

SMITH, F. H., and SMITH, ELIZABETH C., 1942. "Anatomy of the inferior ovary of *Darbya*," *Amer. Jour. Bot.*, **29**: 464–471.

———, ———, 1943. "Floral anatomy of the Santalaceae and some related forms," *Oregon State Monographs, Stud. in Bot.*, **5**: 1–93.

SOUÈGES, R., 1934–1939. *L'embryologie végétale*, Paris: Hermann and Company.

STEBBINS, G. L., 1940. "Studies in the Cichorieae: *Dubyaea* and *Soroseris*, endemics of the Sino-himalayan region," *Mem. Torrey Bot. Club*, **19**(3): 1–76.

SWAMY, B. G. L., 1949. "Further contributions to the morphology of the Degeneriaceae," *Jour. Arnold Arb.*, **30**: 10–38.

———, and BAILEY, I. W., 1950. "*Sarcandra*, a vesselless genus of the Chloranthaceae," *Jour. Arnold Arb.*, **31**: 117–129.

TEPFER, S., 1953. "Floral anatomy and ontogeny in *Aquilegia formosa* var. *truncata* and *Ranunculus repens*," *Univ. Calif. Publ. Bot.*, **25**: 513–648.

TIAGI, Y. D., 1955. "Studies in floral morphology. II. Vascular anatomy of the flower of certain species of Cactaceae," *Jour. Indian Bot. Soc.*, **34**: 408–428.

TSINGER, N. V., 1958. *Semya, ego razvitie i physiologicheskie svoistva (The seed, its development and physiological properties)*, Akademii Nauk SSSR.

WINKLER, H., 1940. "Zur Einigung und Weiterführung in der Frage des Fruchtsystems," *Beitr. Biol. Pflanz.*, **27**: 92–130.

WODEHOUSE, R. P., 1928. "The phylogenetic value of pollen characters," *Ann. Bot.*, **42**: 891–934.

———, 1935. *Pollen grains*, New York: McGraw-Hill.

———, 1936. "Evolution of pollen grains," *Bot. Rev.*, **2**: 67–84.

WUNDERLICH, ROSALIE, 1959. "Zur Frage der Phylogenie der Endospermtypen bei den Angiospermen," *Öst. Bot. Zeit.*, **106**: 203–293.